I: Biological
lliam McD.

Cancer,
sad, *Editors*

Group
dvanced
Philip H.
Editors

Critical
Cynthia S.

, and Cancer:
David

:
Management,
, *Editors*

Research,

Approaches to
ent Diabetes
r R. Diani,

ther Aspects of
Clinical
Pines, Michael
rs

lecular
Haseltine, Neal

ens C. Skou,
bach, Mikael
wo volumes:
B: *Cellular*

y Group
ontroversies in
. Schröder, Jan
bert M. Pinedo,
e Voogt, *Editors*

Membrane
yn E. Howell,
rd Evans,

stoma
Giulio J.
obert C.

Vol 272: **Bacterial Endotoxins: Pathophysiological Effects, Clinical Significance, and Pharmacological Control,** Jack Levin, Harry R. Büller, Jan W. ten Cate, Sander J.H. van Deventer, Augueste Sturk, *Editors*

Vol 273: **The Ion Pumps: Structure, Function, and Regulation,** Wilfred D. Stein, *Editor*

Vol 274: **Oxidases and Related Redox Systems,** Tsoo E. King, Howard S. Mason, Martin Morrison, *Editors*

Vol 275: **Electrophysiology of the Sinoatrial and Atrioventricular Nodes,** Todor N. Mazgalev, Leonard S. Dreifus, Eric L. Michelson, *Editors*

Vol 276: **Prediction of Response to Cancer Therapy,** Thomas C. Hall, *Editor*

Vol 277: **Advances in Urologic Oncology,** Nasser Javadpour, Gerald P. Murphy, *Editors*

Vol 278: **Advances in Cancer Control: Cancer Control Research and the Emergence of the Oncology Product Line,** Paul F. Engstrom, Paul N. Anderson, Lee E. Mortenson, *Editors*

Vol 279: **Basic and Clinical Perspectives of Colorectal Polyps and Cancer,** Glenn Steele, Jr., Randall W. Burt, Sidney J. Winawer, James P. Karr, *Editors*

Vol 280: **Plant Flavonoids in Biology and Medicine II: Biochemical, Cellular, and Medicinal Properties,** Vivian Cody, Elliott Middleton, Jr., Jeffrey B. Harborne, Alain Beretz, *Editors*

Vol 281: **Transplacental Effects on Fetal Health,** Dante G. Scarpelli, George Migaki, *Editors*

Vol 282: **Biological Membranes: Aberrations in Membrane Structure and Function,** Manfred L. Karnovsky, Alexander Leaf, Liana Bolis, *Editors*

Vol 283: **Platelet Membrane Receptors: Molecular Biology, Immunology, Biochemistry, and Pathology,** G.A. Jamieson, *Editor*

Vol 284: **Cellular Factors in Development and Differentiation: Embryos, Teratocarcinomas, and Differentiated Tissues,** Stephen E. Harris, Per-Erik Mansson, *Editors*

Vol 285: **Non-Radiometric Assays: Technology and Application in Polypeptide and Steroid Hormone Detection,** Barry D. Albertson, Florence P. Haseltine, *Editors*

information about previous titles in this series.

PROGRESS IN CLI

Series Editors
Nathan Back V
George J. Brewer F

RECENT TITLES

Vol 235: **Preclinical Safet Products Intended for H** Graham, *Editor*

Vol 236: **First Vienna Sho** Schlag, Heinz Redl, *Editor* volumes: Part A: *Pathophy Mediators and Mediator In* Part B: *Monitoring and Tre*

Vol 237: **The Use of Trans the Diagnosis and Manage Cancer,** Fred Lee, Richard

Vol 238: **Avian Immunolog** D.L. Ewert, *Editors*

Vol 239: **Current Concepts the Study of Prostate Canc** Coffey, Nicholas Bruchovsky Gardner, Jr., Martin I. Resni *Editors*

Vol 240: **Pathophysiological Cell Vaso-Occlusion,** Ronald

Vol 241: **Genetics and Alcoh** Goedde, Dharam P. Agarwal,

Vol 242: **Prostaglandins in C** Helmut Sinzinger, Karsten Sc

Vol 243: **Prostate Cancer,** Ge Saad Khoury, Réne Küss, Chri Louis Denis, *Editors*. Publishe Part A: *Research, Endocrine T Histopathology*. Part B: *Imagin Radiotherapy, Chemotherapy, Issues*

Vol 244: **Cellular Immunother** Robert L. Truitt, Robert P. Gal Bortin, *Editors*

Vol 245: **Regulation and Contr Smooth Muscle,** Marion J. Sieg Somlyo, Newman L. Stephens,

Vol 258: **Membrane Biophysics I Transport,** Mumtaz A. Dinno, W Armstrong, *Editors*

Vol 259: **Nutrition, Growth, and** George P. Tryfiates, Kedar N. Pr

Vol 260: **EORTC Genitourinary Monograph 4: Management of Cancer of Prostate and Bladder,** Smith, Michele Pavone-Macaluso

Vol 261: **Nicotine Replacement: Evaluation,** Ovide F. Pomerleau, Pomerleau, *Editors*

Vol 262: **Hormones, Cell Biolog Perspectives and Potentials,** W. Hankins, David Puett, *Editors*

Vol 263: **Mechanisms in Asthma Pharmacology, Physiology, and** Carol L. Armour, Judith L. Blac

Vol 264: **Perspectives in Shock** Robert F. Bond, *Editor*

Vol 265: **Pathogenesis and New the Study of Noninsulin-Depen Mellitus,** Albert Y. Chang, Arth *Editors*

Vol 266: **Growth Factors and O Wound Healing: Biological and Implications,** Adrian Barbul, El Caldwell, Thomas K. Hunt, *Edit*

Vol 267: **Meiotic Inhibition: M Control of Meiosis,** Florence P. L. First, *Editors*

Vol 268: **The Na^+, K^+-Pump,** J Jens G. Nørby, Arvid B. Mauns Esmann, *Editors*. Published in t Part A: *Molecular Aspects*. Part *Aspects*.

Vol 269: **EORTC Genitourinar Monograph 5: Progress and C Oncological Urology II,** Fritz G.M. Klijn, Karl H. Kurth, He Ted A.W. Splinter, Herman J.

Vol 270: **Cell-Free Analysis of Traffic,** D. James Morré, Kath Geoffrey M.W. Cook, W. How *Editors*

Vol 271: **Advances in Neurobl Research 2,** Audrey E. Evans, D'Angio, Alfred G. Knudson, Seeger, *Editors*

Please contact the publisher fo

CELLULAR FACTORS IN DEVELOPMENT AND DIFFERENTIATION

Embryos, Teratocarcinomas, and Differentiated Tissues

Address all Inquiries to the Publisher
Alan R. Liss, Inc., 41 East 11th Street, New York, NY 10003

Printed in the United States of America

Library of Congress Cataloging-in-Publication Data

International Symposium on Cellular Endocrinology (3rd : 1987 : Lake Placid, N.Y.)

Cellular factors in development and differentiation : embryos, teratocarcinomas, and differentiated tissues : proceedings of the Third International Symposium on Cellular Endocrinology, held at Lake Placid, New York, August 30–September 2, 1987 / editors, Stephen E. Harris, Per-Erik Mansson.

p. cm.—(Progress in clinical and biological research ; v. 284)

Includes bibliographies and index.

ISBN 0-8451-5134-7

1. Cell differentiation—Congresses. 2. Gene expression–Congresses. 3. Mammals—Development—Congresses. I. Harris, Stephen E., 1942- . II. Mansson, Per-Erik. III. Title. IV. Series.

[DNLM: 1. Cell Differentiation—drug effects—congresses. 2. Embryo—drug effects—congresses. 3. Gene Expression Regulation–congresses. 4. Hormones—physiology—congresses. 5. Oncogenes–congresses. W1 PR668E v. 284 / WK 102 I6087c 1987]

QH607.I55 1987

574.87'612—dc19

DNLM/DLC

for Library of Congress 88-22997

CIP

Cellular Factors in Development and Differentiation
Embryos, Teratocarcinomas, and Differentiated Tissues

Proceedings of the Third International Symposium on Cellular Endocrinology
Held at Lake Placid, New York, August 30–September 2, 1987

Editors

Stephen E. Harris
and
Per-Erik Mansson
W. Alton Jones Cell Science Center
Lake Placid, New York

ALAN R. LISS, INC. • NEW YORK

Contents

Contributors

Claire Bonnerot, Department of Molecular Biology, Institut Pasteur, 75015 Paris, France **[125]**

Ralph L. Brinster, Laboratory of Reproductive Physiology, University of Pennsylvania, Philadelphia, PA 19104 **[31]**

Maja Bucan, Imperial Cancer Research Fund, Lincoln's Inn Fields, London WC2A 3PX, England **[147]**

Brent H. Cochran, Department of Biology, Center for Cancer Research, Massachusetts Institute of Technology, Cambridge, MA 02139 **[225]**

David T. Crowe, Department of Cell Biology, Baylor College of Medicine, Houston, TX 77030 **[211]**

Robert Distel, Department of Biological Chemistry and Molecular Pharmacology, Division of Cancer Pharmacology, Harvard Medical School, Dana-Farber Cancer Institute, Boston, MA 02115 **[187]**

Yoshihiro Endo, Division of Reproductive Biology and Endocrinology, University of Pennsylvania, Philadelphia, PA 19104 **[77]**

Mathias Fibi, Department of Molecular Cell Biology, Max Planck Institute of Biophysical Chemistry, 3400 Göttingen, Federal Republic of Germany **[93]**

Walter J. Gehring, Department of Cell Biology, Biozentrum, University of Basel, CH-4056 Basel, Switzerland **[105]**

Peter Gruss, Department of Molecular Cell Biology, Max Planck Institute of Biophysical Chemistry, 3400 Göttingen, Federal Republic of Germany **[93]**

Jonathan Ham, Imperial Cancer Research Fund, London WC2A 3PX, England **[53]**

Robert E. Hammer, Howard Hughes Medical Institute, University of Texas Southwestern Medical Center at Dallas, Dallas, TX 75235-9050 **[1]**

Marie A. Harris, W. Alton Jones Cell Science Center, Lake Placid, NY 12946 **[53]**

Stephen E. Harris, W. Alton Jones Cell Science Center, Lake Placid, NY 12946 **[53]**

The numbers in brackets are the opening page numbers of the contributors' articles.

Timothy E. Hayes, Department of Biology, Center for Cancer Research, Massachusetts Institute of Technology, Cambridge, MA 02139 **[225]**

Joseph Herbert, Department of Neurology, Columbia University College of Physicians and Surgeons, New York, NY 10032 **[163]**

Young-Ping Hwung, Department of Cell Biology, Baylor College of Medicine, Houston, TX 77030 **[211]**

Alexandra Joyner, Department of Anatomy, University of California at San Francisco, San Francisco, CA 94143 **[147]**

Michael Kessel, Department of Molecular Cell Biology, Max Planck Institute of Biophysical Chemistry, 3400 Göttingen, Federal Republic of Germany **[93]**

Gregory S. Kopf, Division of Reproductive Biology and Endocrinology, University of Pennsylvania, Philadelphia, PA 19104 **[77]**

Henry M. Krause, Department of Cell Biology, Biozentrum, University of Basel, CH-4056 Basel, Switzerland **[105]**

Shigeaki Kurasawa, Division of Reproductive Biology and Endocrinology, University of Pennsylvania, Philadelphia, PA 19104 **[77]**

Hans Lehrach, Imperial Cancer Research Fund, Lincoln's Inn Fields, London WC2A 3PX, England **[147]**

Arnold J. Levine, Department of Molecular Biology, Princeton University, Princeton, NJ 08544 **[163]**

Joan Lin, Department of Molecular Biology, Princeton University, Princeton, NJ 08544 **[163]**

Guillermina Lozano, Department of Molecular Biology, Princeton University, Princeton, NJ 08544 **[163]**

Jeffrey R. Marks, Department of Molecular Biology, Princeton University, Princeton, NJ 08544 **[163]**

Gail R. Martin, Department of Anatomy, University of California at San Francisco, San Francisco, CA 94143 **[147]**

Peter Mattei, Department of Biology, University of Pennsylvania, Philadelphia, PA 19104 **[77]**

Douglas C. Miller, Department of Pathology and Neuropathology, New York University Medical Center, New York, NY 10016 **[163]**

Maurice Needham, Imperial Cancer Research Fund, London WC2A 3PX, England **[53]**

Jean-François Nicolas, Department of Molecular Biology, Institut Pasteur, 75015 Paris, France **[125]**

Bert W. O'Malley, Department of Cell Biology, Baylor College of Medicine, Houston, TX 77030 **[37]**

Malcolm G. Parker, Imperial Cancer Research Fund, London WC2A 3PX, England **[53]**

G. Barry Pierce, Department of Pathology, University of Colorado School of Medicine, Denver, CO 80262 **[33]**

Ikuko Sagami, Department of Cell Biology, Baylor College of Medicine, Houston, TX 77030 **[37]**

Richard M. Schultz, Department of Biology, University of Pennsylvania, Philadelphia, PA 19104 **[77]**

Lee M. Silver, Department of Molecular Biology, Princeton University, Princeton, NJ 08540 **[147]**

Bruce M. Spiegelman, Department of Biological Chemistry and Molecular Pharmacology, Division of Cancer Pharmacology, Harvard Medical School, Dana-Farber Cancer Institute, Boston, MA 02115 **[187]**

John Townes, Department of Biology, Center for Cancer Research, Massachusetts Institute of Technology, Cambridge, MA 02139 **[225]**

Ming-Jer Tsai, Department of Cell Biology, Baylor College of Medicine, Houston, TX 77030 **[37,211]**

Sophia Y. Tsai, Department of Cell Biology, Baylor College of Medicine, Houston, TX 77030 **[37,211]**

Heng Wang, Department of Cell Biology, Baylor College of Medicine, Houston, TX 77030 **[37]**

Lee-Ho Wang, Department of Cell Biology, Baylor College of Medicine, Houston, TX 77030 **[37]**

Paul Webb, Imperial Cancer Research Fund, London WC2A 3PX, England **[53]**

Roger White, Imperial Cancer Research Fund, London WC2A 3PX, England **[53]**

Preface

The Third International Symposium on Cellular Endocrinology was held at Lake Placid, New York in September 1987. The meeting was in honor of Dr. Ralph L. Brinster and Dr. Leroy C. Stevens for their pioneering work on mammalian embryogenesis and development of the transgenic animal and the teratocarcinoma model. Their techniques and studies have opened up whole new areas of analysis of a variety of new cellular factors involved in oogenesis, embryogenesis, and terminal differentiation.

After highlighting the many intellectual, spiritual, and technical contributions that Dr. Brinster and Dr. Stevens have made during the many years of their involvement in analysis of developmental problems, several scientific sessions were devoted to subjects of special interest both to the honored guests and to the participants in the symposium. These scientific sessions clearly demonstrated how Drs. Brinster and Stevens have stimulated many scientists to make major contributions to the study of mammalian development.

Areas covered in the symposium include analysis of gene expression in differentiated tissues, factors that play an important role in oocyte maturation and fertilization, and use of transgenic mice to elucidate the basic mechanism of gene programming during development. A session on teratocarcinomas demonstrated how this system has given, and will continue to give, insight into basic biological questions concerning cellular factors, both endogenous and exogenous, that play a role in development. Using transgenic animal models, several new insights into cancer and other aberrant growth patterns are now open to intense investigation. Finally, a session on *cis* DNA elements and *trans*-acting factors during gene activation gave the feeling that complex developmental gene circuits may be dissected in the future by means of in vitro analysis together with transgenic animal systems.

We again thank Dr. Ralph L. Brinster and Dr. Leroy C. Stevens for their contributions toward making it possible for scientists to study transgenic animal systems as the basis for elucidating a variety of fundamental developmental problems.

Stephen E. Harris
Per-Erik Mansson

Symposium participants

Acknowledgments

The Organizing Committee would like to thank the following corporations for their generous contribution to our Third International Symposium on "Cellular Factors in Development and Differentiation—Embryos, Teratocarcinomas and Differentiated Tissues":

Beckman Instruments, Inc.
Boehringer Mannheim Corporation
Bristol-Myers Company
Cetus Corporation
Collaborative Research, Inc.
DNAX Research Institute of Molecular and Cellular Biology, Inc.
E.I. Du Pont De Nemours & Co., Inc.
Genentech, Inc.
Hoffmann-Laroche, Inc.
Johnson & Johnson
Lab-Line Instruments, Inc.
Life Technologies, Inc.
Merck Sharp & Dohm Research Laboratories
Monsanto Company
Sandoz Research Institute
Sulzer Biotech Systems
The Upjohn Company
VWR Scientific

We would also like to acknowledge the valuable skills and aid in the organization of this Symposium from Carla A. Hamelin, Renee Johnston and Julie Lamb. Also, thanks to Marina LaDuke and Polly Butler for their audio/visual assistance.

Due to their efforts our Symposium was a success.

Cellular Factors in Development and Differentiation:
Embryos, Teratocarcinomas, and Differentiated Tissues, pages 1–30

THE SCIENTIFIC CONTRIBUTIONS OF RALPH L. BRINSTER TO UNDERSTANDING MAMMALIAN EMBRYO DEVELOPMENT AND EUKARYOTIC GENE EXPRESSION

Robert E. Hammer

Howard Hughes Medical Institute and Department of Cell Biology, University of Texas Southwestern Medical Center, Dallas, Texas 75235

INTRODUCTION

For more than 25 years, Ralph Brinster has been working and publishing in a broad area loosely termed early embryo development. His interest in this topic began during training in veterinary medicine at the University of Pennsylvania where he became aware of the enormous embryonic mortality associated with the time of implantation. Consequently, when he finished professional school in 1960, he began graduate training in reproductive physiology with the intention of studying preimplantation stages of embryo development. From the beginning his research grew in scope and complexity, contributing in a fundamental manner to a broad area of biology including reproductive processes, embryo development and gene expression. Brinster's contributions, while continuous during the past 25 years, arise from three broad overlapping series of experiments. The first series began in 1961, extended into the 1970s, and dealt with the development of an egg culture system and its utilization to study metabolism. The second series began in 1968, extended into the 1970s, and was directed toward modifying the genotype of a mouse by transferring potential stem cells into the blastocyst cavity. The third series of experiments began about 1976, extend to the present time, and involves the introduction of macromolecules, including foreign genes, into the early embryo to study the regulation of gene expression. It is clear that these three areas of investigation have a common central theme of embryo development, but because distinct seminal contributions have been made in each area, they will be reviewed separately.

EGG CULTURE AND METABOLISM

In 1961, when Brinster began his studies on early mammalian embryo development, there was not a reliable in vitro culture system to investigate the preimplantation stages (for review see Brinster, 1964; 1972). Although little was known about the early embryo of any species, the mouse appeared to be the most promising model to study because the genetic background was the best understood, and the small size of the mouse provided obvious experimental and economic advantages.

Hammond (1949) had successfully cultured 8-cell mouse eggs to blastocysts by employing a salt solution containing glucose and egg-white. Subsequently, Whitten (1956, 1957) added lactate to a Krebs-Ringer bicarbonate solution and obtained cleavage of 2-cell mouse eggs to blastocysts. From this background, Brinster began a series of experiments to develop a repeatable, convenient culture technique to evaluate critically the metabolic requirements of the embryo. After testing more than a dozen different types of culture systems, he found that a plastic Petri dish containing "microdrops" of culture medium under paraffin oil allowed consistent, in vitro development of 2-cell mouse eggs into blastocysts (Brinster, 1963; 1964). The technique was simple and permitted observation of the cells at any time during culture (Figure 1). The advantages of this technique have become increasingly apparent over the last 25 years, and it has been adopted by most workers that perform experiments on early embryos of all mammalian species.

Brinster employed this technique to analyze in detail the metabolic requirements of the preimplantation mouse embryo from the 2-cell to blastocyst stage (Brinster, 1965a,b,c,d). In a series of experiments from 1963 to 1967, he defined a basic culture medium for the 2-cell mouse egg (reviewed in Brinster, 1969; 1972). The main characteristics of the medium were a modified Krebs-Ringer bicarbonate salt solution base, supplemented with pyruvate, lactate, bovine serum albumin, and glucose (Table 1). The most fundamental and critical finding was

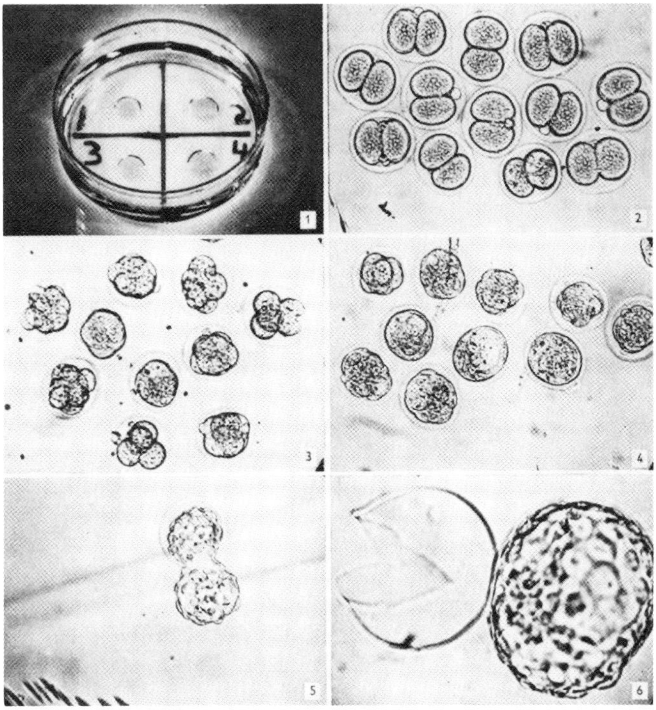

Figure 1. In vitro culture of mouse ova (Brinster, 1963). (1) Culture dish with four microdrops of media under paraffin oil. (2) Two-cell mouse ova at beginning of culture. (3) Eight-cell stage. (4) Morulae and early blastocysts. (5) Late blastocyst escaping from zona pellucida. (6) Late blastocyst free of zona.

that the 2-cell mouse egg required pyruvate as an energy source, unlike other cells grown in vitro which rely on glucose (Brinster, 1965b). A need for pyruvate by the one-cell mouse egg and oocyte was subsequently demonstrated (Biggers et al., 1967; Whittingham and Biggers, 1967). Likewise, early primordial germ cells in the mouse fetal gonadal ridge were shown to rely on pyruvate as an energy source (Brinster and Harstad, 1977). Thus, primordial germ cells, oocytes and early egg cleavage stages of the mouse depend on pyruvate as their central energy source suggesting the possibility

Table 1. Brinster's Medium for Ovum Culture

Component	g/1	mM
NaCl	5.546	94.88
Sodium Lactate	2.253	25.00
Sodium Pyruvate	0.028	0.25
KCl	0.356	4.78
$CaCl_2$	0.189	1.71
KH_2PO_4	0.162	1.19
$MgSO_4 \cdot 7H_2O$	0.294	1.19
$NaHCO_3$	2.106	25.00
Penicillin	100 U/ml	--
Streptomycin	50 μg/ml	--
Bovine Serum Albumin	1.0	--

Brinster, 1965b; 1972.

that this requirement may be an inherent characteristic of germ-line cells.

The development of a reproducible culture technique allowed a systematic examination of energy metabolism in the egg. To establish a base for these metabolic studies, both the oxygen consumption (Mills and Brinster, 1967) and protein content (Brinster, 1967a) of the mouse egg were determined. These determinations permitted quantitative comparisons to be made between cleavage stage embryos and somatic tissues, and these values still form a base of reference for current metabolic work on the egg. Brinster's first metabolic studies in eggs focused on the utilization of various radioactive substrates. In a typical experiment, shown in Table 2, he demonstrated that in the early stages of mouse development, pyruvate can account for nearly 100 per cent of the oxygen consumed by the egg, whereas glucose can only account for a small percentage of oxygen used. Employing similar radioactive substrate techniques it was possible to confirm and extend the metabolic requirements first observed in culture experiments (reviewed in Brinster 1969; 1972). Thus, Brinster established that the requirement for pyruvate was a characteristic of the eggs of many if not all mammalian species, suggesting a unique metabolism in these cells (Brinster, 1969; 1972).

Table 2. Comparison of Carbon Dioxide Production From Glucose and Pyruvate By Mouse Eggs

Stage of Development	CO_2 from Glucose	CO_2 from Pyruvate
Unfertilized	0.13±0.01 (2)	7.24±0.11 (104)
Fertilized	0.68±0.02 (10)	6.95±0.79 (99)
Two-cell	1.19±0.09 (18)	6.03±0.18 (90)
Eight-cell	2.16±0.08 (25)	7.25±0.54 (85)
Morula	6.73±0.69 (42)	10.00±0.86 (64)
Blastocyst	10.94±1.11 (53)	13.92±1.23 (68)
Late Blastocyst	14.69±0.77 (62)	15.73±2.27 (66)

Values are pico moles ± SEM formed per embryo per hour. In parentheses is the per cent of total oxygen consumed at that stage represented by the CO_2 (Brinster, 1967b; 1969).

The activity of enzymes regulating energy metabolism was determined in cleavage stage embryos in order to help identify possible key regulatory reactions. Particularly surprising was the high level of lactic dehydrogenase (LDH) activity in the eggs of many species (Brinster, 1965e; 1973). In the mouse, LDH represents approximately 5% of the protein content of the one-cell eggs. The most important enzyme measurements were those related to glycolysis. Brinster demonstrated that hexokinase, the first enzyme in the metabolic reactions for glucose utilization, had a very low activity early in cleavage. By comparing substrate utilization and enzyme activity, he demonstrated that this enzyme was a key regulator of glucose utilization in cleavage stage mouse eggs (Brinster, 1973).

While more extensive studies were being conducted on energy metabolism of the egg, Brinster and his students initiated a series of key experiments on establishing a profile of egg protein and nucleic acid metabolism. Having developed microassays to quantitate egg protein content, similar techniques were utilized to study protein synthesis and decay in the egg. Brinster demonstrated that during the preimplantation stages of egg development in the mouse, protein content decreases while the rate of protein synthesis increases. This occurs as a result of an increase in protein turn-over rate (Brinster et al., 1976). Furthermore, individual proteins in the egg have vastly different half-lives (Brinster et al., 1979). In his studies on mouse egg nucleic acids, Brinster and his students were among the first to demonstrate that α-amanitin, a selective inhibitor of RNA polymerase II, was a potent inhibitor of development from the 1-cell stage onward (Levey et al., 1977). They went on to document an increase in poly A containing RNA (putative messenger RNA) from the 2-cell to blastocyst stage (Levey et al., 1978). Their studies suggested an early and critical role for transcription in mammalian eggs.

Among the most important contributions from this series of experiments on egg development were the elaboration of a simple, reliable egg culture system and the identification of pyruvate as the central energy source for mouse eggs, as well as eggs of other species. These two contributions, subsequently extended by Brinster and by others, have provided the means to carry out much of the exciting work that is now being conducted on early embryos. "The elucidation of adequate in vitro culture conditions in the 1960s probably stands out as the single most important advance in the field" (Papaioannou, 1987).

TERATOCARCINOMA CELL TRANSFER INTO BLASTOCYSTS

While studying egg culture and metabolism in the mid 1960s, Brinster became interested in factors that regulated early mammalian egg development and subsequent differentiation. He considered that one approach to studying such events would be to modify the genotype of the embryo and study the subsequent effects on phenotype. Tarkowski (1961) and Mintz (1962) had demonstrated that the production of chimeras by fusing 8-cell embryos was an extremely valuable technique for studying developmental questions. Subsequently, Gardner (1968) reported that the inner cell mass cells of the blastocyst, the cells destined to become the embryo, could be transferred into another blastocyst, survive, and take part in development. Chimeras formed by this technique could be used to investigate systematically the development of distinct cell lineages during early differentiation. Brinster reasoned that other types of nonembryonic "stem cells" transferred into the blastocyst might colonize the embryo and allow the study of their contribution to the adult animal. To pilot this approach a series of experiments were undertaken beginning in 1968 in which cells from the mouse blastocyst inner cell mass and cells from cultured blastocysts were transferred into recipient mouse blastocysts by microinjection (Figure 2). These studies were designed to develop the technique of cell transfer and to determine if older embryonic cells could colonize the blastocyst and participate in development. While transfer of synchronous cells resulted in the development of chimeras, cells from cultured blastocysts only a few days older than the recipient blastocyst (day 4) had very little if any chance to colonize (Moustafa and Brinster; 1972a;b). However, the experimental usefulness of a system that would allow nonembryonic cells, particularly cells that could be cultured and modified in vitro, to colonize and participate in embryo development seemed so great that experiments in this area were continued.

Beginning in 1970, Brinster began stem cell transfer experiments that involved two cell types, bone marrow cells and teratocarcinoma cells. In the bone marrow stem cell experiments, the transferred cells were obtained from the femur of CBA T6T6 animals and injected into day 4 blastocysts. About 80 of the offspring which developed

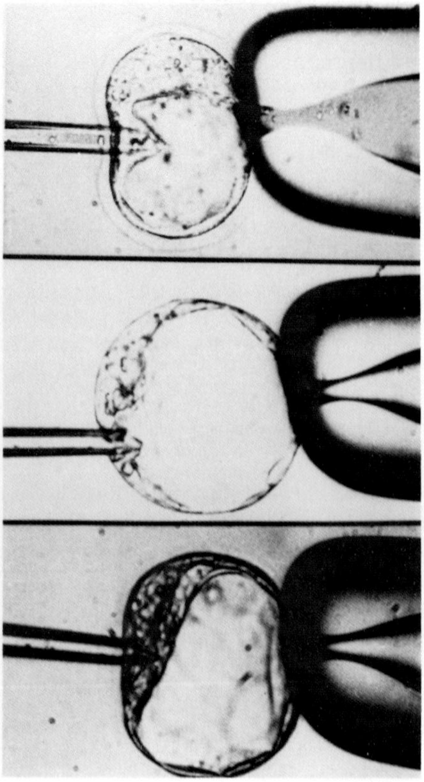

Figure 2. Penetration of mouse blastocysts by transfer pipettes. (a) trophoectoderm layer of a recipient blastocyst resisting a blunt tip penetration (b) adhesion of transferred cells to the inner cell mass of the recipient blastocyst (c) donor cells may be transplanted directly into the inner cell mass (Moustafa and Brinster, 1972a).

from these injected blastocysts received skin grafts from the donor cell mouse strain, and surprisingly a significant increase in graft survival time occurred compared to control animals (Brinster, 1974). The results indicated that the injected cells survived and influenced the immune response of the developing fetus. In the second set of experiments, testicular teratocarcinoma cells (OTT 6050) were harvested from the ascites fluid of 129 SvSl mice that are used to carry this tumor. The mice carrying the testicular teratocarcinoma cells were a generous gift of Roy Stevens who had isolated the original tumor and adapted it to grow as embryoid bodies in ascites fluid (Stevens, 1964; and elsewhere in this volume). The embryoid body tumor masses resemble normal mouse embryos of the late morula and blastocyst stage. However, they proliferate as a carcinoma and will kill a mouse in approximately three weeks following intraperitoneal injection. Cells from these masses were isolated and injected into blastocysts from Swiss albino random bred mice. A single teratocarcinoma cell is capable of giving rise to an individual tumor that can contain many differentiated tissues (Kleinsmith and Pierce, 1964). The injected blastocysts were then introduced into pseudopregnant females to continue development. The general scheme of the experiment is shown in Figure 3. Sixty of the progeny from this

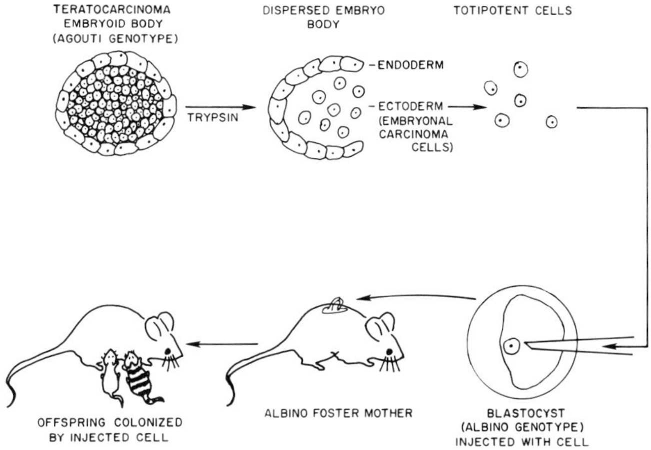

Figure 3. Technique for the introduction of teratocarcinoma cells into mice.

experiment were analyzed by skin grafting and again a significantly prolonged graft survival time was found (Brinster 1974). However, the most dramatic finding was the presence of several thin stripes of agouti hair in one of the animals arising from an injected blastocyst (Figure 4). The injected cells (OTT 6050) had come originally from an agouti animal, and the dark stripes clearly represented the participation of transferred cells in development of at least the dermis of the mouse. In addition, the malignant cells had been maintained for many generations in the laboratory as an ascites teratocarcinoma. Thus, the malignancy of cells that normally would kill a mouse in three weeks could be

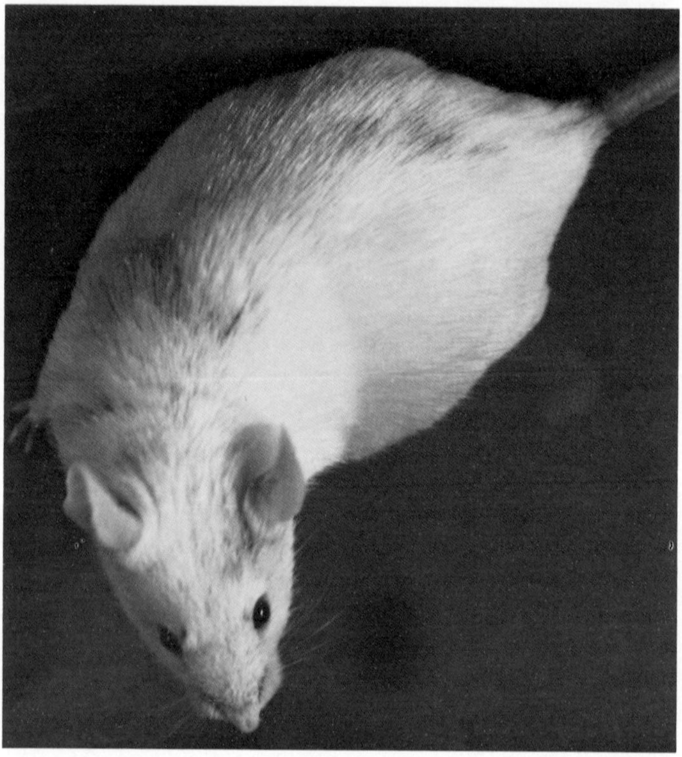

Figure 4. A random-bred Swiss albino mouse with several stripes of agouti hair extending from the midline, right flank and thorax. The animal developed from a blastocyst of Swiss albino genotype that received 129 SvSl (agouti) teratocarcinoma cells (Brinster, 1974).

controlled by the blastocyst, and could take part in development. In the following year, these experiments were confirmed and greatly extended in two other laboratories (Mintz and Illmensee, 1975; Papaioannou, et al., 1975).

The demonstration that teratocarcinoma cells could colonize the embryo and participate in development was important for several reasons aside from the intriguing finding that the blastocyst could block the malignant proliferation of the cells. First, it was demonstrated that these cells could be grown in vitro and used for a variety of studies on differentiation (Martin and Evans, 1975; Sherman and Solter, 1975). Consequently, their participation in normal development lent creditability to in vitro studies in which these cells were used as models to study the process of differentiation. Second, the ability to introduce a cell with a modified genotype into a mouse and study the resulting consequences of its presence, one of Brinster's original objectives, seemed feasible. One could conceivable mutate the cells in vitro, introduce them into the mouse, and if they colonized the germ-line, the mutation should be recoverable in the progeny. This ideal approach was not realized for several reasons, particularly the efficiency of colonization was low; germ-line participation was infrequent; and the cells became aneuploid when grown in vitro (Dewey et al., 1977; Russell, 1978). However, recent advances have caused a considerable resurgence of interest in the transfer of stem cells into the blastocysts. Alternative cell lines have now been derived from outgrowths of normal blastocysts. These have been designated Evans/Kaufman (EK) cells (Evans and Kaufman, 1981) or embryonal stem (ES) cells (Martin et al., 1981). These cells colonize blastocysts efficiently, frequently participate in formation of the germ-line, and remain euploid in culture, thereby allowing selection. Consequently, a variety of experiments have been performed employing this technique, including the production of specific models of human disease (Kuehn et al., 1987; Hooper et al., 1987). Thus the simple cell transfer experiments initiated almost 20 years ago in Brinster's laboratory sparked the interest of other groups who contributed in many ways to a system that is now increasingly important for studies on development and gene function.

MACROMOLECULE INJECTION INTO EGGS

During the late 1960s, Brinster worked on egg nuclear transfer and egg cell fusion as a means of modifying the mouse genome. However, manipulated eggs never developed more than a few division, and therefore, alternative approaches were considered. McBride and Ozer (1973) reported that genetic information could be transferred into cultured cells by purified metaphase chromosomes, and the possibility of performing similar experiments with mouse eggs seemed exciting. Consequently, in 1976 and 1977, when the teratocarcinoma cell transfer experiments were progressing slowly because of difficulty with maintaining euploid cell lines, Brinster decided to undertake chromosome transfers. Clearly obtaining pure and biologically active preparations of chromosomes was a problem. However, the field of recombinant DNA had been developing rapidly, and it had been demonstrated that purified genes were expressed following transfection into cultured cells (Graham and van der Eb, 1973; Wigler et al., 1977). Furthermore, Jaenisch and Mintz (1974) had found that viral DNA injected into the blastocyst cavity was present in the cells of the adult.

Encouraged by the availability of pure preparations of genes, pilot experiments were begun in 1977 to inject macromolecules, both mRNA and DNA , into mouse eggs. The plan was to begin with messenger RNA and progress to genes transcribed by RNA polymerase type III and then genes transcribed by RNA polymerase type II, an approach that had been used to study the Xenopus egg (Gurdon, 1977). Activity of the injected mRNA and/or genes could be measured following culture of the eggs. The final and most important experiments would involve the transfer of eggs injected with cloned genes into pseudopregnant females and analysis of the resulting animals for incorporation of the injected genes.

In the first series of experiments, the cytoplasm of fertilized mouse eggs was injected with mRNA for β-globin prepared from mouse or rabbit reticulocytes. Following injection, the eggs were incubated in ^{3}H-amino acids and the labelled proteins separated by two-dimensional gel eletrophoresis (Brinster et al., 1979; 1980a). Fluorography clearly indicated the synthesis of the appropriate globin protein, and a typical example is shown in Figure 5. These studies were extended to examine synthesis of other proteins from injected mRNA as well as to determine the half-life of both the new protein and mRNA. Thus, these studies provided

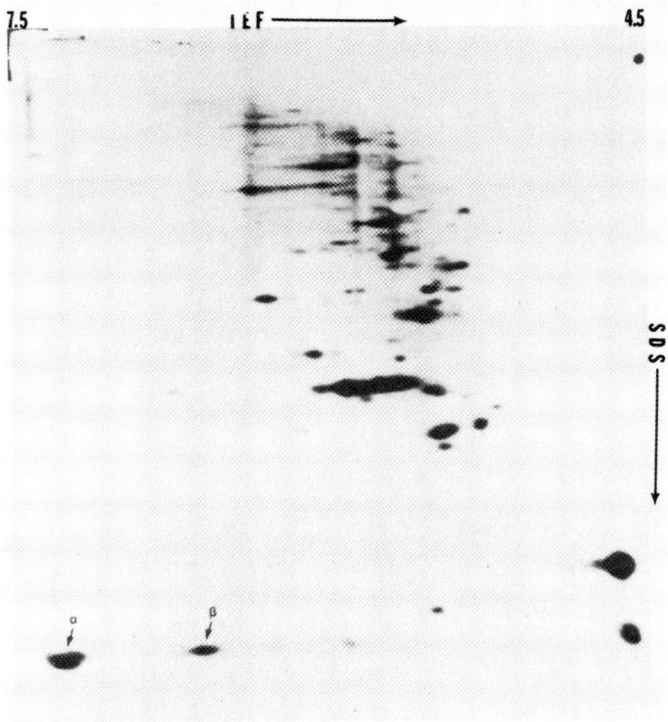

Figure 5. Fluorograph of two-dimensional electrophoretic separation of proteins from one-cell fertilized ova injected with mouse globin mRNA (Brinster et al., 1980a). The two radioactive spots indicated by the arrows coincide with the position of the non-radioactive globin marker protein prepared from mouse blood. α and β indicate the position of α- and β-chains of globin.

information about mRNA turnover and protein metabolism in the egg and served as pilot studies for future DNA experiments.

In the next set of studies, the Xenopus 5S RNA gene was injected into the germinal vesicle of the mouse oocyte and transcripts produced from this gene detected electrophoretically with a 5S RNA marker, an indication that the gene was transcribed and processed accurately (Brinster et al., 1980b; 1981a). Injections of the same gene into the pronucleus of the fertilized egg did not result in detectable levels of 5S RNA, indicating significant differences between these stages of egg development in the transcription of polymerase III genes and probably similar endogenous genes.

Most eukaryotic genes are transcribed by RNA polymerase II, and therefore, a logical next step was to inject such genes into mouse eggs. The herpes simplex virus thymidine kinase (HSV-TK) gene was selected for preliminary studies and injected into growing oocytes and fertilized eggs (Brinster et al., 1981b). In contrast to the result with polymerase III genes, both the oocyte and fertilized egg transcribed the HSV gene and translated the message formed into functional viral thymidine kinase (see Table 3). This transient assay technique has proven to be a rapid and useful method for defining the critical regulatory regions of genes such as metallothionein (Brinster et al., 1982; Stuart et al., 1984). Furthermore, the results with the HSV-TK gene demonstrated an enormous increase (30-fold) in the TK levels found in the fertilized egg compared to mature oocytes (Chen et al., 1986). This dramatic increase in transcriptional ability is one of the earliest and most important events in the program of activation following fertilization.

These studies with messenger RNA, polymerase II genes, and polymerase III genes, were experiments in which results were obtained quickly since eggs were assayed within 24 to 48 hours following injection. In addition to the information they provided about the ability of the egg to process RNA and DNA, these experiments established techniques necessary to introduce genes into animals.

Table 3. Herpes Simplex Virus Thymidine Kinase Activity in Early Stages of Mouse Ova Injected With the Cloned Gene for the Enzyme

Stage of Development	Site of Injection[a]	Plasmid Injected[b]	Formation of $[^3H]TMP$ (cpm x 10^{-3})
Growing Oocytes	-	None	2.3 ± 0.3[c]
	GV	pBR322	2.7 ± 0.3
	GV	pTK	34.0 ± 2.4
Mature Oocytes	-	None	6.1 ± 0.6
	GV	pTK	55.1 ± 17.6
Unfertilized Ova	-	None	6.1 ± 0.8
	C	pTK	9.6 ± 2.8
Fertilized	-	None	8.7 ± 1.2
	PN	pBR322	6.7 ± 1.0
	PN	pTK	1,580.0 ± 150.0
	C	pTK	8.4 ± 0.7

a GV, germinal vesicle; C, cytoplasm; PN, pronucleus.

b The plasmids injected were pBR322; pTK, which contains the herpes simplex viral thymidine kinase gene in pBR322.

c Following injection, the ova were cultured for 22 hours and $[^3H]$ thymidine monophosphate ($[^3H]TMP$), the reaction product of thymidine kinase, was then assayed (Brinster et al., 1982). Values are for 25 ova and are mean ± SEM for three or more experiments. (Brinster et al., 1981b; Chen et al., 1986).

DEVELOPMENT OF TRANSGENIC MICE

In 1979 and 1980, while the experiments described above were underway, five different genes, including the rabbit β-globin gene, SV 40 DNA, an HSV-TK construct, a sea urchin histone gene and the Xenopus 5S RNA gene were injected into fertilized eggs which were transferred to recipient females, and animals developing from these eggs examined for gene incorporation. In only one case was there tentative evidence for gene integration (RLB, personal communication). Early in 1980, Brinster contacted Richard Palmiter to obtain ovalbumin mRNA for an RNA microinjection experiment. At that time Palmiter was working on characterizing the regulatory region of the mouse metallothionein-1 (MT) gene and Brinster was assaying eggs for expression of the HSV-TK gene and analyzing livers of mice which developed from injected eggs for evidence of TK expression. To facilitate delineation of the MT 5' flanking DNA in transfection experiments, Palmiter fused the MT promoter and 5' flanking DNA to a convenient reporter gene, the HSV-TK structural gene. Brinster and Palmiter then decided to test the same construct in mouse eggs as a convenient alternative transient expression assay system. At the same time, MT-TK injected eggs were transferred into pseudopregnant females and allowed to develop to term. These collaborative studies began in January 1981 and resulted in a critical delineation of the MT promoter as well as mice that integrated the gene into their chromosomes (Brinster et al., 1981c). With these exciting results began a close and continuing collaboration between Brinster and Palmiter in developing and utilizing transgenic animals to investigate eukaryotic gene expression.

Between December 1980 and November 1981, five papers appeared describing the results of experiments in which DNA was injected into the pronucleus of mouse eggs and became integrated into chromosomes (Gordon et al., 1980; E. Wagner et al., 1981; T. Wagner et al., 1981; Costantini and Lacy, 1981; Brinster et al., 1981c). In two studies, no gene expression was reported (Gordon et al., 1980; Costantini and Lacy, 1981), and in another, low level expression of the foreign gene was seen in an animal (E. Wagner et al., 1981). T. Wagner et al., (1981) reported that rabbit globin was found in the blood of some adult mice that had developed from eggs microinjected with the rabbit β-globin gene. In the experiments with the MT-TK gene, high levels of viral thymidine kinase activity were found in the liver of one mouse that contained the gene (Table 4). This mouse (23-2) had about 200-fold more TK activity than control mice. Mice and other animals into which foreign DNA has integrated are called transgenic (Gordon and Ruddle, 1981).

An interesting aspect regarding transgenic animals has been the precocious development of the technique in mice. Frequently, innovative or new experimental techniques in developmental biology are first devised and perfected in lower species, such as Xenopus or Drosophila, where the eggs are larger and the embryos more tractable. However, in this instance, transgenic mice were made before similar success was obtained in other species. Table 5 shows a partial chronology of the development of transgenic animals and plants.

There are basically three key elements in the development of the transgenic mouse methodology. These are: (1) integration of the DNA into chromosomes, (2) high levels of expression of the foreign gene, and (3) tissue-specific expression of an appropriate construct. As indicated above, the first of these elements was clearly established independently in five laboratories by injecting DNA into the pronucleus of fertilized eggs (Table 5).

The second key element in the transgenic mouse methodology, expression of the injected gene, was observed in three of the original reports (E. Wagner et al., 1981; T. Wagner et al., 1981; Brinster et al., 1981c) at variable levels. The TK activity arising from the MT-TK gene was clearly quite high and unequivocal. Results from the experiments using the MT-TK fusion gene were very convincing and encouraging to Brinster and Palmiter. As a result, several other studies using fusion gene constructs were undertaken. One of these was an attempt to correct a genetic defect. The dwarf little _(lit/lit)_ mouse served as an ideal candidate, because this phenotype was thought to be due solely to a deficiency of growth hormone (Eicher and Beamer, 1975; Beamer and Eicher, 1976). On the basis of the success

Table 4. Herpes Thymidine Kinase Gene Expression in Liver of Mice Developing from Eggs Injected with the MT-TK Fusion Gene

Mouse	MT-TK DNA	TK Activity[a]		HSV-TK mRNA (mol/cell)
		-Ab	+Ab	
14-1	-	355	261	< 2
14-2	-	552	425	< 2
19-1	-	593	723	< 2
19-2	+	920	915	7.5
19-3	-	492	368	< 2
21-1	-	656	525	< 2
21-2	-	716	550	< 2
21-3	+	468	579	< 2
21-4	-	770	654	< 2
21-5	-	633	571	< 2
23-1	+	560	546	< 2
23-2	+	23,475	9,416	28
		(121,830)	(4,141)[b]	

a TK activity was measured with either 10 µl of anti-HSV-TK IgG fraction (+Ab) or normal IgG fraction (-Ab). Activity is expressed as cpm ^{3}H-TMP/mg wet wt/min.

b Values in parentheses were obtained when the homogenate from mouse 23-2 was diluted 10-fold prior to assay. For details see Brinster, 1981c.

Table 5. Production of Transgenic Organism

Organism	Gene	Reference
Mice	*SV40 - blastocyst cavity*	Jaenish & Mintz, 1974
	TK, SV40	Gordon et al., 1980
	hβ-globin, TK	E. Wagner et al., 1981
	rβ-globin	T. Wagner et al., 1981
	rβ-globin	Costantini & Lacy, 1981
	mMT-TK	Brinster et al., 1981
Xenopus	*rβ-globin*	Rusconi & Shaffner, 1981
Drosophila	*P element*	Spradling & Rubin, 1982
	XDH	Rubin & Spradling, 1982
Plants	*NOS-NEO*	DeBlock et al., 1984
	NOS-NEO	Horsch et al., 1984
C. elegans	*col-β-gluc.*	Stinchcomb et al., 1985
Sea Urchin	*actin - NEO*	McMahon et al., 1985
Farm Animals	*mMT-hGH*	Hammer et al., 1985

Abbreviations: **species;** h, human; r, rabbit; m, mouse; **genes;** SV40, Simian virus 40 DNA; TK, herpes simplex viral thymidine kinase; MT, metallothionein-1; XDH, xanthine dehydrogenase; NOS, nopaline synthase; NEO, neomycin phosphotransferase; col, collagen; β-gluc., β-glucuroni- dase; GH, growth hormone.

with the MT-TK construct, a metallothionein-rat growth hormone (MT-GH) fusion gene was made. Because breeding sufficient homozygous little mice to provide eggs would take many months, a control experiment was performed in which the MT-GH construct was injected into eggs from normal mice. The results of this experiment were dramatic, seven of 21 mice born from injected eggs contained the gene and six grew faster and became larger than control littermates (Palmiter et al., 1982). One of the mice that expressed the gene was mated and a genetic line established. All the progeny that contained the gene grew larger than controls. A typical growth curve is shown in Figure 6. The impact of these results was

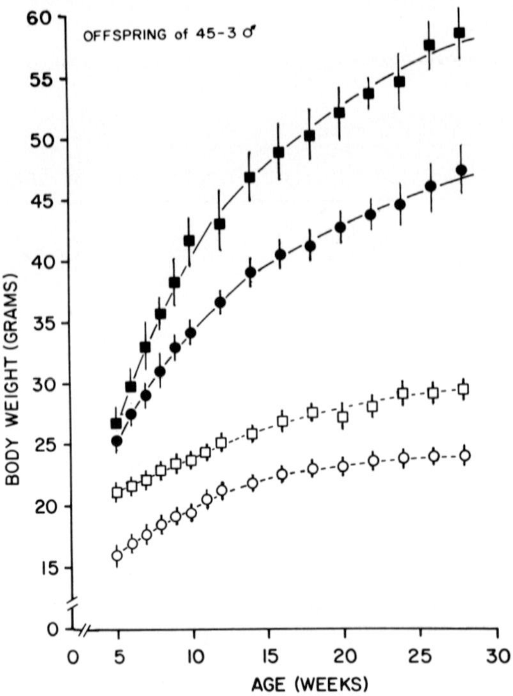

Figure 6. Comparative growth of transgenic (solid symbols) and control (open symbols) progeny of founder mouse 45-3, expressing an MT-rat growth hormone fusion gene. Transgenic mice (males - squares, females - circles) are larger than controls at five weeks of age, enter an accelerated growth phase between five and eleven weeks and by 17 weeks are approximately double normal size (Hammer et al., 1984b).

enormous, and dramatized the potential of transgenic mice. This report conclusively demonstrated high level expression of transgenes and the phenotypic effect of foreign gene expression was self-evident. These experiments captured the interest and imagination of the general public. The experiment to correct a genetic defect by germ-line therapy was completed the following year with the transferred MT-GH gene enhancing the growth of the dwarf little mice (Hammer et al., 1984a).

The third key element in establishing the transgenic system was the demonstration that injected genes would function in a tissue-specific manner. That this might be true was suggested by the results from experiments with the MT-TK and MT-GH fusion genes as well as the chicken transferrin gene (McKnight et al., 1983). However, the first clear demonstration of tissue-specific expression of a transgene was provided by a rearranged kappa light chain immunoglobulin gene (Brinster, et al., 1983; Storb et al., 1984). Six different kappa transgenic mouse lines were produced, and in each case the injected gene was expressed in a tissue-specific manner (Figure 7). The immunoglobulin genes represent a special class of genes since they undergo rearrangement before expression. However, the tissue-specific expression of microinjected genes that are not rearranged was also demonstrated using the rat elastase gene (Swift et al., 1984). Levels of mRNA from the injected elastase gene were shown to be as high or higher than found in the acinar cells of the rat exocrine pancreas (Figure 8). These studies with the elastase gene unequivocally demonstrated the tissue-specific expression of microinjected genes.

The documentation of these three elements was essential to establishing the transgenic technique as a powerful and widely useful method in biology. Brinster and Palmiter were one of the groups demonstrating DNA integration in 1981; thus establishing the first element. Their growth hormone experiment, probably the best known experiment in transgenics, was extremely important in providing evidence for effective transgene expression, and thus contributed greatly to establishing the second element important to the methodology. Finally, in collaboration with the Storb and MacDonald laboratories,

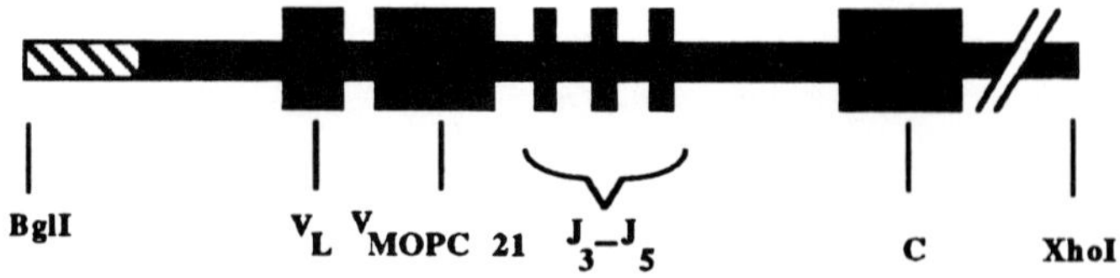

Animal	Spleen mRNA: pg κRNA per μg total RNA	Spleen mRNA: %κ M.21
47	254	32
48	4*	39
49	27	40
52	17	25
54	27	28
55	96	26
Control	19	2

Figure 7. Tissue-specific expression of a rearranged immunoglobulin κ gene in transgenic mice (Brinster et al., 1983; Storb et al., 1984). The κ gene was derived from the myeloma MOPC-21 (M.21). No κ M.21 mRNA was detectable in other tissues of founder animals. The injected κ gene construct had 1.3 and 7.8 kb of 5' and 3' DNA respectively and 0.9 kb of pBR322 (crosshatched).

* This value is artificially low due to partial degradation of spleen RNA.

Brinster provided the first conclusive evidence for tissue-specific expression of microinjected genes. There have been many other contributions by Brinster and Palmiter to transgenic methodology, and the use of this technique to study gene regulation. A complete review of this field may be found in Brinster and Palmiter (1986) or Palmiter and Brinster (1986).

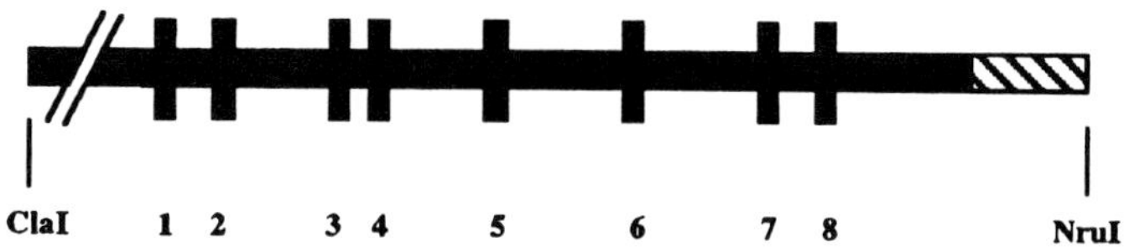

Animal	Pancreas mRNA (molecules/cell)
6-3	45,000
9-4	18,000
13-1	120,000
13-4	10,000
22-5-1	34,000
Rat	10,000

Figure 8. Tissue-specific expression of the rat elastase gene in transgenic mice (Swift et al., 1984). The elastase gene (solid regions) had 7 kb of 5' flanking DNA and about 1kb of pBR322 DNA (crosshatched). Expression of rat elastase in other tissues was less than 40 molecules per cell.

SUMMARY AND KEY PUBLICATIONS

It is difficult to summarize in a few paragraphs research contributions spanning 25 years. One can only hope to mention the most important. A simple way to achieve such a brief overview of an individual might be to list their most important publications and indicate how each contributed to science. There will be, of course, some differences of opinion among scientists making such a list, but from Brinster's approximately 200 publications, I think the following ten are likely to be considered highly significant by many workers in the fields of egg culture, embryo development, and gene regulation.

1. Brinster, R.L. (1963) Exp. Cell Res. 32:205. This publication first described the basic technique that has since been adopted by most investigators to culture mammalian eggs of all species. The introduction of this technique and the demonstration of pyruvate as the principal energy source (see below) greatly facilitated subsequent studies on mammalian embryo development.

2. Brinster, R.L. (1965) J. Exp. Zool. 158:59. This publication first identified pyruvate rather than glucose as the central energy source of mouse eggs. It was later shown by Brinster's group and by others to be applicable to all early mouse egg stages and to the eggs of other species.

3. Brinster, R.L. (1974) J. Exp. Med. 140:1049. This publication showed that a teratocarcinoma cell could be injected into a mouse blastocyst and the descendants of the cell would colonize tissues of the animal developing from the blastocyst. It demonstrated the feasibility of this approach to modify the genome of animals.

4. Brinster, R.L., Chen, H.Y., Trumbauer, M.E., Avarbock, M.R. (1980) Nature 283:499. This publication reports the accurate translation of microinjected foreign mRNA by the mouse egg. It indicates the feasibility of introducing macromolecules to examine their processing in the egg.

5. Brinster, R.L. Chen, H.Y, Trumbauer, M.E. (1981) Science 211:396. This publication reports the transcription of a microinjected foreign gene by the egg and indicates the potential use of this system for understanding gene function in eggs. Publications 4 and 5 represent fundamental experiments in the evolution of Brinster's progressive development of transgenic experiments in his laboratory.

6. Brinster, R.L., Chen, H.Y., Trumbauer, M.E., Senear, A.W., Warren, R. and Palmiter, R.D. (1981c) Cell 27:223. This publication describes the integration and expression of the MT-TK fusion gene in mice. It is among the first reports of transgenic mice

and, because of the high level of gene expression, indicates the enormous possibilities of the technique.

7. Palmiter, R.D., Brinster, R.L., Hammer, R.E., Trumbauer, M.E., Rosenfeld, M.G., Birnberg, N.C. and Evan, R.M. (1982) Nature 300:611. This publication describes the production of "giant" mice following introduction of a foreign growth hormone gene. It had a powerful impact on the scientific community and is an important milestone in science.

8. Brinster, R.L., Ritchie, K.A., Hammer, R.E., O'Brien, R.L., Arp, B. and Storb, U. (1983) Nature 306:332. This publication was the first to document tissue-specific expression in transgenic mice, which was critical to establishing the full power of the system.

9. Brinster, R.L., Chen, H.Y., Messing, A., Van Dyke, T., Levine, A.J. and Palmiter, R.D. (1984) Cell 37:367-379. This publication is the first description of oncogenesis in mice resulting from the introduction of a transgene. It demonstrates the feasibility of obtaining unique models of cancer and studying the mechanism of oncogenesis in this system.

10. Brinster, R.L., Chen, H.Y., Trumbauer, M.E., Yagle, M.K. and Palmiter, R.D. (1985) PNAS 82:4438. This publication describes many of the critical factors necessary for efficient introduction of DNA into mice and reflects the enormous number of transgenic mice produced (nearly 3000) by the Brinster/Palmiter team, probably several times greater than all other laboratories combined up to that time.

This small group of publications reflects the high caliber and seminal nature of the contributions made by Brinster and his colleagues to the field of embryo development and gene expression over the past 25 years. First, his pioneering work in establishing egg culture methods and characterizing embryo metabolism set a foundation on which much of the research on mammalian embryo development has depended. Second, his experiments demonstrating that teratocarcinoma cells transferred into the mouse blastocyst cavity colonize the developing animal showed that the mouse genotype and hence phenotype could be modified by this approach. Third, his contributions to development of transgenic methodology and particularly his contributions to understanding the regulation of gene expression, the basis of development, have been extensive and exceptional. Throughout his scientific career, Brinster has continued to provide and utilize innovative ways to investigate development. Furthermore, his contributions have sparked the interest of investigators from diverse disciplines to study events controlling early mammalian development and gene expression. Ralph Brinster clearly has been one of the most influential scientists in mammalian developmental biology.

ACKNOWLEDGEMENTS

This presentation was given to recognize and honor the enormous contribution Ralph Brinster has made during the past 25 years to the field of mammalian developmental biology. I had the opportunity and good fortune to be a postdoctoral fellow in Ralph's laboratory from 1982 to 1986. During this time I came to appreciate the significance of Ralph's early contributions to the fields of embryo culture, embryo metabolism and embryo manipulation, all of which paved the way for his most recent studies on gene expression. Having spent many hours with Ralph, I experienced firsthand his relentless and systematic pursuit of investigations. Ralph's publications are a record of novel insights into complex questions and reflect his unyielding determination and tenacity.

The importance of Ralph's scientific contributions have been recognized by numerous groups. In 1975, he was appointed Richard King Mellon Professor of Reproductive Physiology at the University of Pennsylvania School of Veterinary Medicine. He received, with Richard Palmiter, the New York Academy of Science Award in Biology and Medical Sciences in 1983. In 1984, he was invited to give a Harvey Society Lecture, and in 1986 he was elected to the Institute of Medicine of the National Academy of Sciences and became a Fellow of the American Academy of Arts and Sciences. This past year he was elected to the

National Academy of Sciences. It is, indeed, an honor for me to have had this opportunity to highlight Ralph's career. He has been more than a mentor, he remains a close friend and has provided a continual source of inspiration and encouragement.

I thank Lori Baker for her assistance in the preparation of this manuscript.

REFERENCES

Beamer WG, Eicher EM (1976). J Endocr 71:37.
Biggers JD, Whittingham DG, Donahue RP (1967). Proc Natl Acad Sci 58:560.
Brinster RL (1963). Exp Cell Res 32:205.
Brinster RL (1964). Ph.D. Thesis. University of Pennsylvania.
Brinster RL (1965a). J Exp Zool 158:49.
Brinster RL (1965b). J Exp Zool 158:59.
Brinster RL (1965c). J Exp Zool 158:69.
Brinster RL (1965d). J Reprod Fert 10:227.
Brinster RL (1965e). Biochim Biophys Acta 110:439.
Brinster RL (1967a). J Reprod Fert 13:413.
Brinster RL (1967b). Exp Cell Res 47:634.
Brinster RL (1969). In: The Mammalian Oviduct. Hafez ESE, Blandau R (eds), Chicago: University of Chicago Press, p 419.
Brinster RL (1972). In: Growth, Nutrition, and Metabolism of Cells in Culture. Rothblat G, Cristofalo V (eds), New York: Academic Press, Inc., p 251.
Brinster RL (1973). In: Regulation of Mammalian Reproduction. Segal SJ et al (eds), Springfield: Charles C. Thomas Publishing Company, p 302.
Brinster RL (1974). J Exp Med 104:1049.
Brinster RL, Brunner S, Joseph X, Levey IL (1979). J Biol Chem 254:1927.
Brinster RL, Chen HY, Avarbock MR, Trumbauer ME (1979). J Cell Biol 83:32A.
Brinster RL, Chen HY, Messing A, Van Dyke T, Levine AJ, Palmiter RD (1984). Cell 37:367.
Brinster RL, Chen HY, Trumbauer ME (1980b). J Cell Biol 87:283A.
Brinster RL, Chen HY, Trumbauer ME (1981a). Science 211:396.

Brinster RL, Chen HY, Trumbauer ME (1981b). J Cell Biol 91:364A.
Brinster RL, Chen HY, Trumbauer ME, Avarbock MR (1980a). Nature 283:499.
Brinster RL, Chen HY, Trumbauer ME, Senear AW, Warren R, Palmiter RD (1981c). Cell 27:223.
Brinster RL, Chen HY, Trumbauer ME, Yagle MK, Palmiter RD (1985). Proc Natl Acad Sci 82:4438.
Brinster RL, Chen HY, Warren R, Sarthy A, Palmiter RD (1982). Nature 296:39.
Brinster RL, Harstad H (1977). Exp Cell Res 109:111.
Brinster RL, Palmiter RD (1986). In: Harvey Lectures Series 80. New York: Alan R. Liss, Inc. p 1.
Brinster RL, Ritchie KA, Hammer RE, O'Brien RL, Arp B, Storb U (1983). Nature 306:332.
Brinster RL, Wiebold JL, Brunner S (1976). Dev Biol 51:215.
Chen HY, Trumbauer ME, Ebert KM, Palmiter RD, Brinster RL (1986). In: Molecular Developmental Biology. Bogorad L (ed). New York: Alan R. Liss, p 149.
Costantini F, Lacy E (1981). Nature 294:92.
DeBlock M, Herrera-Estrella L, Van Montagu, Schell J, Zambryski P (1984). EMBO J 3:1681.
Dewey MJ, Martin DW Jr, Martin GR (1977). Proc Natl Acad Sci 74:5564.
Eicher EM, Beamer WG (1976). J of Heredity 67:87.
Evans MJ, Kaufman MH (1981), Nature 292:154.
Gardner RL (1968). Nature 220:596.
Gordon JW, Ruddle FH (1981). Science 214:1244.
Gordon JW, Scangos GA, Plotkin DJ, Barbosa JA, Ruddle FH (1980). Proc Natl Acad Sci 77:7380.
Graham FL, van der Eb AJ (1973). Virology 52:456.
Gurdon JB (1977). Proc R Soc Lond B 198:211.
Hammer RE, Palmiter RD, Brinster RL (1984a). Nature 311:65.
Hammer RE, Palmiter RD, Brinster RL (1984b). Labrie F, Proulx, L (eds) In: Endocrinology. Elsevier Science, p 651.
Hammer RE, Pursel VG, Rexroad CE Jr., Wall RJ, Bolt DJ, Ebert KM, Palmiter RD, Brinster RL (1985). Nature 315:680.
Hammond J, Jr. (1949). Nature 163:28.
Hooper M, Hardy K, Handyside A, Hunter S, Monk M (1987). Nature 326:292.
Horsch R, Fraley RT, Rogers SG, Sanders PR, Lloyd A, Hoffmann N (1984). Science 223:496.

Jaenisch R, Mintz B (1974). Proc Natl Acad Sci 71:1250.
Kleinsmith LJ, Pierce GB Jr (1964). Can Res 24:1544.
Kuehn MR, Bradley A, Robertson EJ, Evans MJ (1987). Nature 326:295.
Levey IL, Stull GB, Brinster RL (1978). Dev Biol 64:140.
Levey IL, Troike DE, Brinster RL (1977). J Reprod Fert 50:147.
Martin GR (1981). Proc Natl Acad Sci 78:7634.
Martin GR, Evans MJ (1975). Cell 6:467.
McBride OW, Ozer HL (1973). Proc Natl Acad Sci 70:1258.
McKnight GS, Hammer RE, Kuenzel EA, Brinster RL (1983). Cell 34:335.
McMahon AP, Flytzanis, CN, Hough-Evans BR, Katula KS, Britten RJ, Davidson EH (1985). Dev Biol 108:420.
Mills RM Jr, Brinster RL (1967). Exp Cell Res 47:337.
Mintz B (1962). Am Zool 2:432, (abstr).
Mintz B, Illmensee K (1975). Proc Natl Acad Sci 72:3585.
Moustafa LA, Brinster RL (1972a). J Exp Zool 181:181.
Moustafa LA, Brinster RL (1972b). J Exp Zool 181:193.
Palmiter RD, Brinster RL (1986). 1986 Annu Rev Genet 20:465.
Palmiter RD, Brinster RL, Hammer RE, Trumbauer ME, Rosenfeld MG, Birnberg NC, Evans RM (1982). Nature 300:611.
Papaioannou VE (1987). Cell 49:159.
Papaioannou VE, McBurney MW, Gardner RL, Evans MJ (1975). Nature 258:70.
Rubin GM, Spradling AC (1982). Science 218:348.
Rusconi S, Shaffner W (1981). Proc Natl Acad Sci 78:5051
Russell LB (ed) (1978). Genetic Mosaics and Chimeras in Mammals. New York: Plenum Press.
Sherman MI, Solter D (1975). Teratomas and Differentiation. New York: Academic Press.
Spradling AC, Rubin GM (1982). Science 218:341.
Stevens LC (1964). Proc Natl Acad Sci 52:654.
Stinchcomb DT, Shaw J, Carr S, Hirsh D (1985). In: Genetic Manipulation of the Early Mammalian Embryo. Cold Spring Harbor Lab., p123.
Storb U, O'Brien RL, McMullen MD, Gollahon KA, Brinster RL (1984). Nature 310:238.
Stuart GW, Searle PF, Chen HY, Brinster RL, Palmiter RD (1984). Proc Natl Acad Sci 81:7318.
Swift GH, Hammer RE, MacDonald RJ, Brinster RL (1984). Cell 38:639.
Tarkowski, AK (1961). Nature 190:857.

Wagner EF, Stewart TA, Mintz B (1981). Proc Natl Acad Sci 78:5016.
Wagner TE, Hoppe PC, Jollick JD, Scholl DR, Hodinka RL, Gault JB (1981). Proc Natl Acad Sci 78:6376.
Whitten WK (1956). Nature 176:96.
Whitten WK (1957). Nature 179:1081.
Whittingham D, Biggers JD (1967). Nature 213:942.
Wigler M, Silverstein S, Lee LS, Pellicer A, Cheng YC, Axel R (1977). Cell 11:223.

Cellular Factors in Development and Differentiation:
Embryos, Teratocarcinomas, and Differentiated Tissues, pages 31–32

COMMENT

R. L. Brinster

School of Veterinary Medicine
University of Pennsylvania
Philadelphia, Pennsylvania 19104

To have one's research recognized by a symposium is an extraordinary honor, and it is indeed a pleasure to share this honor with such a distinguished scientist as Roy Stevens who has contributed so much to the study of differentiation. I am grateful to the W. Alton Jones Cell Science Center for this recognition and thank the many talented scientists who participated in the Symposium and contributed chapters to the published volume.

During the course of a scientific career, assistance comes from many sources, and it gives me great pleasure to thank my friends and colleagues in science who have shared their thoughts and collaborated with me on numerous occasions. My interaction with Richard Palmiter has been particularly enjoyable and fruitful. In addition, I especially thank the many individuals who have worked with me in the laboratory as assistants, students and colleagues. It is their enormous energy and talent that have made experiments successful over the years.

I am particularly grateful to Bob Hammer who has given such an authoritative overview of my research. He has been one of those exceptional students and colleagues who bring great satisfaction and reward to a mentor and friend.

Finally, I must say how pleased I am to have been able to contribute to understanding how the embryo develops. The process by which a single cell, the fertilized egg, can evolve into a complex animal through the cell-specific, temporal activation of a gene program is the most fascinating, complicated and important subject man will ever study. Much has been accomplished in the last 25 years, but our understanding is just beginning. Certainly a bright future lies ahead in this field.

Cellular Factors in Development and Differentiation:
Embryos, Teratocarcinomas, and Differentiated Tissues, pages 33–35

Leroy C. Stevens: An Appreciation

G. Barry Pierce, M.D.

Department of Pathology, University of Colorado School of Medicine, Denver Colorado 80262

This meeting honors Leroy Stevens and Ralph Brinster, and I am delighted and honored by the invitation to make a few remarks about Dr. Stevens, a close friend for over 30 years. When one thinks about Roy, one thinks of teratomas (who else has personalized license plates "129TER"), innovative science, generosity, and kindness.

1942 was a memorable year for this native New Yorker. He graduated in Biology from Cornell, married the lovely Jean Davis, and joined the Army. He served in Africa, Sicily, Italy, and landed on the Normandy beaches on D Day. Although he was decorated twice, he prefers to talk about his record at OTC where he claims to have graduated with more demerits than any other candidate.

After the war, Roy earned the Ph.D. with J. Holtfretter, and is therefore a scientific descendant of Hans Spemann. Not a bad pedigree! Upon graduation in 1952 he accepted a position with C.C. Little at the Jackson Laboratory where he has remained since. Roy examined inbred strains of mice for evidences of inherited disease: one day he found a testicular tumor in a strain 129 mouse. More tumors were found, which were transplantable and closely resembled human testicular teratomas and teratocarcinomas. They formed the basis of the three original articles on testicular tumors in strain 129 mice (with C.C. Little, Proceedings of National Academy of Science USA 40:1080-87, 1954; with K.P. Hummel J. Natl. Cancer Institute 18:719-47, 1957; and J. Natl. Cancer Institute 20:1257-75, 1958). I commend these papers to you, because they set the stage for most of the work that has been done on differentiation and cancer. I remember the early meetings when Roy presented his observations and the intensive questioning by Hans Schlumberger, Harry Green, and Jacob Furth, made for great excitement and intellectual stimulation.

Roy attempted to modulate the incidence of teratomas with hormones and chemicals, but the only thing that significantly increased the incidence was incorporation of the steel gene onto the 129 background. He plotted the incidence of occurrence of tumors with time and by extrapolation decided that the teratocarcinogenic event must have occurred precisely 12½ days post-fertilization. Transplantation of genital ridges from 12½ day old strain 129 embryos gave a high incidence of testicular teratocarcinoma identifiable seven days after transplantation. On the other hand, transplantation of genital ridges lacking primordial germ cells failed to produce tumors. Later Roy produced teratomas and teratocarcinomas by transplanting early embryos and traced the origin of the tumors to undifferentiated cells, probably the equivalent of the inner cell mass cells. Then came the LT strain of mouse, ova of which undergo parthenogenetic development to implantation stages before the parthenotes die. He found that parthenogenetic cells could be rescued by including them in normal blastocysts. These are only a few of Roy Stevens' magnificent accomplishments, and it is important to note that as he developed new tumors and new techniques for producing tumors he was unselfish in distributing these tools to whomever wished them. He was kind to all, but especially to young investigators and he had the patience to share his vast knowledge with them.

But there is more to Roy Stevens than the scientist, he and Jean have three wonderful children and five grandchildren. One of my fond remembrances is seeing a picture of Roy teaching his three year old grandson to split wood while sitting down. We wish Roy and Jean every best wish for a happy and productive retirement.

Cellular Factors in Development and Differentiation:
Embryos, Teratocarcinomas, and Differentiated Tissues, pages 37–51

The Studies Of COUP Transcription Factor From HeLa Cells: Purification and Partial Characterization

Lee-Ho Wang, Sophia Y. Tsai, Heng Wang, Ikuko Sagami, Ming-Jer Tsai and Bert W. O'Malley*

Department of Cell Biology, Baylor College of Medicine, Houston, Texas 77030

SUMMARY

A transcription factor COUP, which is responsible for the recognition of the Chicken Ovalbumin Upstream Promoter sequence between -70 and -90, is required for efficient transcription of the ovalbumin gene. This COUP transcription factor has been purified ~200,000-fold by a combination of conventional column and sequence specific DNA affinity column chromatography. A few polypeptides were identified in the purified preparation on an SDS-gel. Upon renaturation, all the major polypeptides in the molecular size range between 43 to 53 kD bound specifically to the COUP sequence. Furthermore, at least one of the renatured polypeptides in the region of 43-45 kD retained transcriptional activity. Using band-shifting assays and 3' and 5' deletion mutants and oligonucleotide mutants, the sequence important for binding was mapped. Methylation and ethylation interference assays revealed the contact sites for the purine base and the phosphate backbone. These studies indicate that the COUP transcription factor interacts with the base residue in the major groove and wraps around the DNA helix in one full turn.

INTRODUCTION

Transcriptional activities of RNA polymerase II is mediated by a complex array of cis-acting elements that enhance or repress the gene expression. Recently, a number of transacting factors which bind specifically to these cis-acting elements have been identified. For instance, Spl which binds to the GC box sequence (Dyan *et al.*, 1986; Jones and Tjian, 1985). The heat shock transcription factor (HSTF) binds to the *Drosophila* heat-shock genes (Parker and Topol, 1984; Topol *et al.*, 1985). The major late transcription factor (MLTF) binds to Ad2-major late promoter (Sawadogo and Roeder, 1985; Carthew *et al.*, 1985). The CCAAT transcription factor (CTF) binds to β-globin, α-globin and TK promoters (Jones *et al.*, 1987; Graves *et al.*, 1986; Cohen *et al.*, 1986). Among these trans-acting factors, Spl (Briggs *et al.*, 1986), CTF (Jones *et al.*, 1987), HSTF (Wiederrecht *et al.*, 1987) and MLTF (Chodosh *et al.*, 1986) have been purified to apparent homogeneity.

For the chicken ovalbumin gene, we have recently defined a cis-acting element which is essential for efficient transcription of the ovalbumin gene (Knoll *et al.*, 1983; Pastorcic *et al.*, 1986). We designated this upstream element, which includes the sequence GTGTCAAAGGTCAAA, as the Chicken Ovalbumin Upstream Promoter (COUP) element. Exonuclease III and DNase I footprinting experiments revealed the presence of a transacting factor(s) in HeLa cell extracts which binds specifically to the COUP sequence (Elbrecht *et al.*, 1985; Sagami *et al.*, 1986). This COUP binding protein stimulated the transcription of the ovalbumin gene *in vitro*.

To understand the detailed mechanism by which the COUP binder regulates the expression of the ovalbumin gene, we set out to purify and characterize the COUP binding protein in more detail. For this purpose we used a specific DNA affinity column containing the concatemerized COUP binding site to purify the COUP binder. In addition, we used methylation and ethylaion interference experiments to define the contact sites.

EXPERIMENTAL PROCEDURES

Preparation of DNA Affinity Resin

Two synthetic oligonucleotides having 19 nucleotides of complementary sequence and four nucleotide cohesive end 5'-TTCTATGGTGTCAAAGGTCAAAC-3' and 5'-AGAAGTTTGACCTTTGACACCAT-3' were separately purified on a 15% polyacrylamide gel and annealed in 0.2 X SSC. The double stranded oligonucleotides were 5'-phosphorylated and ligated to form DNA concatemers. The ligated DNA was covalently attached to CNBr-activated Sepharose 4B as described by Arndt-Jovin *et al.* (1975) and Kadonaga and Tjian (1986). The amount of DNA coupled to a given amount of resin was determined by radioactivity in the preparation. The efficiency of DNA coupling to the resin was more than 80%.

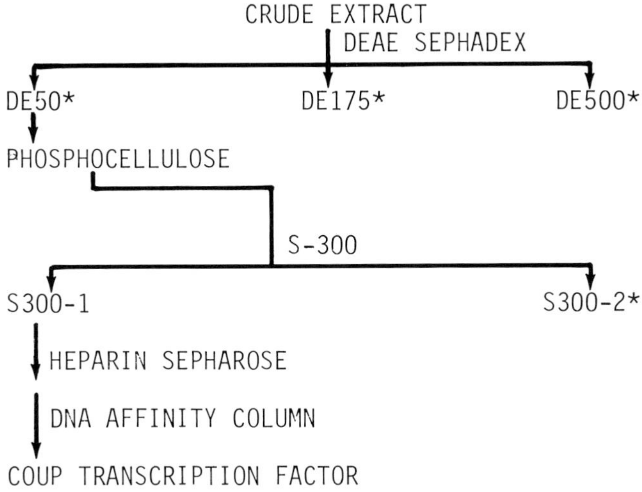

FIGURE 1. Purification scheme of COUP transcription factor from nuclear extract of HeLa cells. Details are given in Experimental Procedures. * represents the fractions which are required for *in vitro* transcription of ovalbumin gene.

Purification of COUP Transcription Factor

Nuclear extracts from HeLa cells were prepared and fractionated by DEAE Sephadex column chromatography as described previously (Pastorcic *et al.*, 1986; Tsai *et al.*, 1981). The flow-through fractions were pooled, dialyzed against 200 mM NaCl in buffer B [20 mM Hepes (pH 7.9), 20% glycerol 0.1 mM EDTA, and 2mM dithiothreitol] and loaded on a 25 ml phosphocellulose column (1.8 x 10 cm). The column was eluted with a linear gradient of NaCl (0.2 to 0.8M) in buffer B. Proteins eluted between 350 and 450 mM NaCl containing COUP binding and transcriptional activity were pooled and applied on a Sephacryl S300 column as described previously (Sagami et al., 1986). The active fractions were pooled and applied on a 5 ml heparin sepharose column (1.1 x 5.5 cm). The column was washed with 12 ml of buffer B + 0.25 M NaCl and eluted with a linear gradient of NaCl (0.25 - 0.6 M NaCl) in buffer B. Fractions which eluted between 320-450 mM NaCl were pooled, and concentrated by reloading on a 0.5 ml heparin sepharose column, step-eluting with buffer B + 0.5 M NaCl. The concentrated fractions were dialyzed against buffer B + 0.1 M NaCl.

DNA Affinity Column Chromatography

The concentrated fractions from the heparin sepharose column chromatography (3 mg protein/1.4 ml) were mixed with excess HinfI - digested pBR322 fragments (1.8 mg/ml) for 30 min at 4°C and then loaded on a DNA affinity column (0.8 x 1.2 cm). The column was then washed extensively with buffer B containing 0.3M NaCl (20 column volumes). The COUP transcription factor was finally eluted with buffer B containing 0.6M NaCl.

Renaturation of SDS-gel Purified COUP Transcription Factor

The DNA affinity column-purified COUP transcription factor was subjected to electrophoresis in an 8% SDS polyacrylamide gel at 4°C, and 5mm gel slices spanning the molecular weight range of 35,000 to 75,000 were excised and eluted electrophoretically at 250 V at 4°C for 4 hours into a dialyzing bag in 0.5 ml of SDS-PAGE buffer containing 100 μg/ml BSA. The gel eluate was precipitated with acetone (acetone:eluate = 4:1) in an ethanol-dry ice bath. The precipitate was collected by centrifugation at 10,000 xg for 20 min as described by Hager and Burgess (1980). The pellets thus obtained were washed with 80% acetone, dried and dissolved in 50 μl of 6 M guanidine-HCl. The mixtures were incubated at room temperature for 20 minutes, and dialyzed overnight against buffer A (Buffer B plus 5 mM $MgCl_2$) containing

100 mM KCl at 4° to allow for renaturation. The renatured samples were then tested for COUP binding activity and *in vitro* transcription activity.

DNA Binding Assays

DNA binding was monitored by band-shifting assays (Fried and Crother, 1981). The conditions of gel electrophoresis and binding were as described previously (Sagami *et al*., 1986). The -44 and -103 fragments designate DNA fragments of -269 to -44 and -269 to -103 flanking the ovalbumin gene, respectively.

Methylation and ethylation interference experiments

The [^{32}P]labelled -44 DNA fragment was methylated of ethylated as described by Sakonju and Brown (1982).

RESULTS

Purification of COUP Transcription Factor

We have identified previously a transcription factor which binds specifically to the COUP sequence of the ovalbumin gene. This COUP transcription factor was purified ~300-fold using conventional column chromatographic techniques such as DEAE-Sephadex, phosphocellulose, Sephacryl-300, and heparin sepharose as diagrammed in Fig. 1. Further steps employing hydroxyapatite, chromatofocusing or hydrophobic column chromatography accomplished little additional purification and recovery was low. For this reason we used a specific DNA affinity column to purify the COUP transcription factor. The amount of DNA covalently attached to the resin was ~50-100 μg per ml resin. It is worth noting that residual ATP from the phosphorylation - ligation reaction should be removed from concatemers before coupling to the CNBr-activated sepharose. Otherwise, the amino group of ATP can link to the resin and significantly lower the efficiency of coupling.

As expected, the majority of the proteins from the heparin sepharose pooled fractions flowed through the DNA affinity column at 0.1 M NaCl in buffer B. The COUP transcription factor was eluted with 0.6 M NaCl in buffer B. The recovery of the binding activity for COUP sequences was about 80%. The amount of COUP factor as estimated by band-shifting assays, assuming a

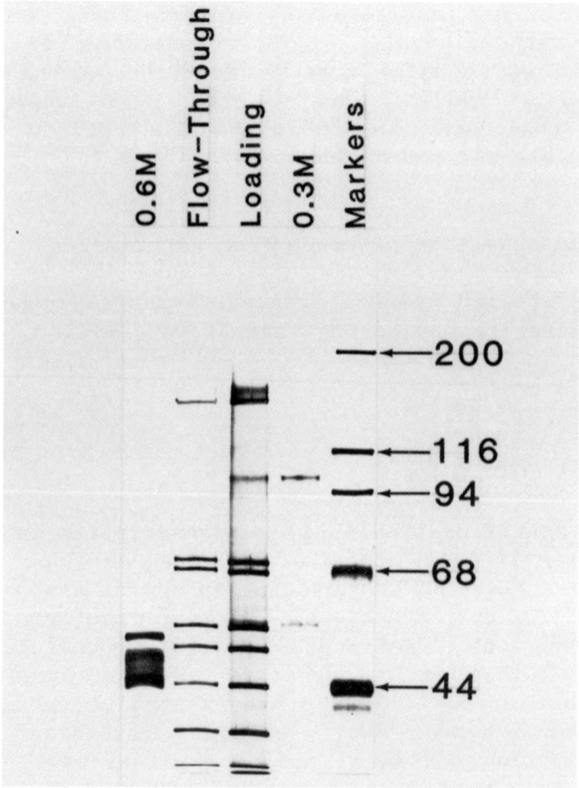

FIGURE 2. SDS-PAGE gel electrophoresis of the purified COUP transcription factor. 0.25 µl out of 400 µl of loading and flowthrough, and 200 µl/800 µ of the 0.6M NaCl-eluted fraction from DNA affinity column were subjected to 8% SDS-PAGE and stained with silver. The 0.6M NaCl fraction was concentrated by precipitation in 10% trichlorocetic acid, then washed with acetone before loading on the gel.

stoichiometry of one molecular protein per DNA binding site, correlates well with the amounts of 43-53 kD polypeptides present in SDS-PAGE gel (determined by comparison of known amounts of standard proteins on a silver-stained gel).

As expected, the affinity-column purified COUP transcription factor could bind to the -44 ovalbumin gene fragment (-44 to -269 upstream of the ovalbumin gene) but not the -103 fragment (-103 to -269) which lacks COUP sequence (data not shown). This data confirmed that the protein recognized the specific sequence. Furthermore, we have determined that the purified COUP transcription factor, when added to an *in vitro* reconstituted transcription reaction, is able to stimulate the transcription of the ovalbumin gene. In the absence of COUP transcription factor, only a low basal level of transcription was observed. However, the addition of affinity-column purified COUP protein resulted in a significant increase in transcription from the ovalbumin promoter (data not shown). Thus, the purified fraction could bind specifically to the COUP sequence and was functional also in an *in vitro* transcription assay.

In order to ascertain which of the polypeptides possessed binding and transcriptional activities, COUP transcription factor purified through the DNA affinity column was separated by SDS-PAGE, 5mm slices were excised and the polypeptides were electroeluted. The SDS was removed by acetone precipitation and the precipitated polypeptides were resuspended in 6 M guanidine hydrochloride followed by dialysis to renature the protein. An aliquot from each sample was tested for binding activity. As shown in Fig. 3, the proteins eluted from slices 3 to 5, corresponding to the molecular size range of 43-53 kD, are capable of binding to DNA fragments containing the COUP sequence. Slice 3 covered the 53 kD polypeptide, slice 4 contained the 48 kD polypeptide and slice 5 included the 46.5, 45 and 43 kD polypeptides (bottom panel of Fig. 3). All of the renatured samples from slice 3 to 6 retained binding activities, but the 43-45 kD fraction displayed the highest level of activity. The low-binding activity detected in slice 6 resulted possibly from the presence of some 43 kD polypeptide, since other experiments showed no binding activity from gel slices below the 43 kD size.

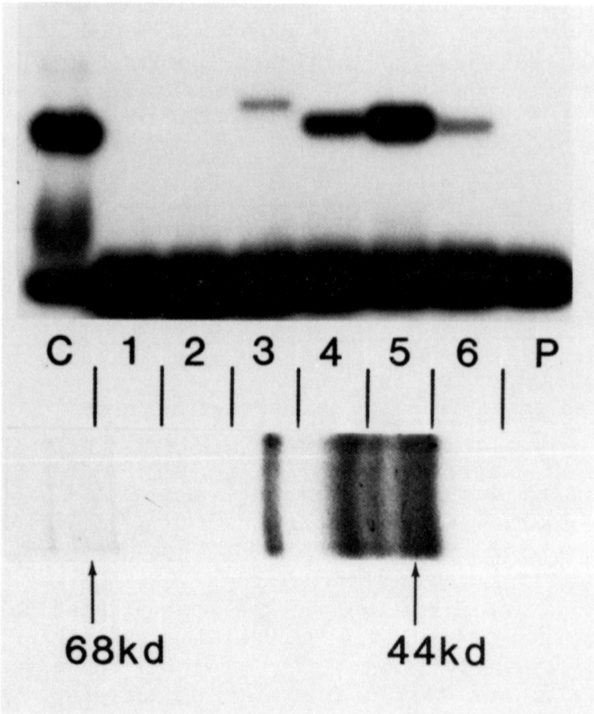

FIGURE 3. Renaturation of SDS-PAGE purified COUP transcription factor. An aliquot of affinity column-purified protein was loaded on 3 lanes of a 8% SDS-PAGE gel. After electrophoresis, the gel was cut into 5 mm slices across 2.5 lanes. The lane which was partially cut was stained by silver to visualize (bottom panel). The remaining two lanes of gel were of renatured samples were monitored by band-shifting assays. C: the control protein sample from affinity column-purified fractions, the number represents the gel slices and P: DNA probe alone.

To further demonstrate the binding specificity, band-shifting competition assays were carried out using

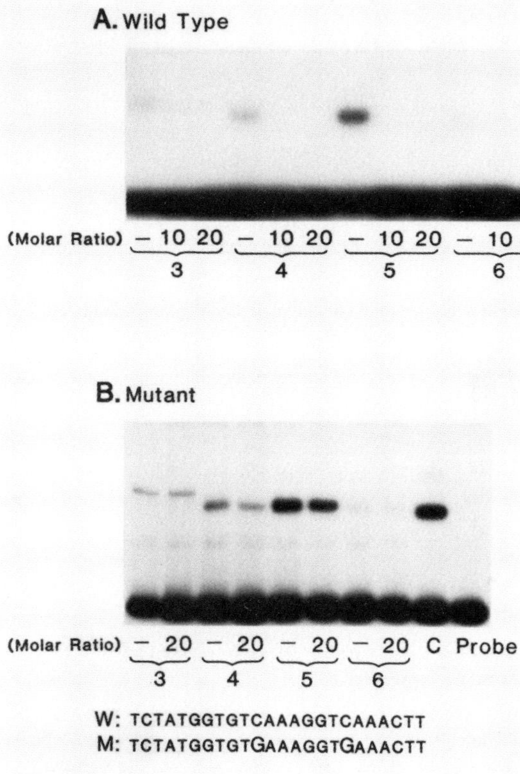

Figure 4a and 4B. Binding specificty of renatured COUP transcription factor. The renatured COUP binding activity was competed by A) wild type and B) mutant oligonucleotides. Renatured samples from slices 3-6 were incubated with labelled -44 fragment wtih increasing amount of unlabelled oligonucleotides. Minus signs indicate no competitor in the mixture. The bottom panel shows the sequences of the wild type and the mutant oligonucleotides.

unlabelled synthetic oligonucleotides, either wild type or mutated sequences as competitor. The wild type double stranded 28-mer corresponds to -91 to -64 of the ovalbumin promoter, while the mutant has a C to G transversion both at positions -81 and -74. The binding of renatured protein(s) from each slice (slices 3 to 6 in Fig. 3) was totally abolished in the presence of 10- and 20-fold molar excess of wild type oligonucleotide (Fig. 4A). In contrast, the mutated oligonucleotide showed no effect on COUP binding even at a molar ratios of 20 (Fig. 4B). Such experiments indicate that the purified polypeptides of 43-53 kD in size are capable of binding selectively to the COUP box sequences of the ovalbumin promoter. Although the recovery of the binding activity from renatured samples is generally poor (~3-5%), it provides a specific means to identify the protein which is the active binding component. To ensure that the renatured samples still retained functional activity, renatured samples from slice 5 containing the highest level of binding activity was used in our _in vitro_ reconstituted transcription system. Addition of the SDS gel eluted and renatured protein enhances the level of transcription of the ovalbumin gene (data not shown). Therefore, we concluded that at least one of 43-45 kD polypeptides was responsible for both the promoter-specific binding and transcription-enhancing activities of the ovalbumin gene.

Binding of the COUP Transcription Factor to Various Ovalbumin Gene Promoter Mutants

As demonstrated earlier (Sagami _et al._, 1986), the COUP transcription factor protects a region which contains two direct repeat sequences of GTCAAA between -71 and -83 of the ovalbumin promoter. To examine more closely the binding site of the COUP transcription factor, we carried out band-shifting assays using different deletion mutants. The COUP transcription factor binds well to 3' deletion mutants having a 3' end point mapped at -75 nucleotides upstream from the RNA start site. However, it fails to bind to 3' deletion fragments terminating at -89 nucleotides upstream from the RNA start site. Similar studies on 5' deletion mutants showed that fragments containing 143 nucleotides of 5' flanking sequence bound COUP factor well, while deletion mutants which contain only 77 or 24 nucleotides of 5' flanking sequence failed to bind.

We concluded that when part of the downstream GTCAAA repeat (3' deletion, -75 fragment) was deleted, it did not affect binding. On the other hand, deletion of the upstream GTCAAA repeat abolished completely the binding (5' deletion, -77 fragment). These results suggest that the upstream GTCAAA repeat is more important for binding and the presence of an intact direct repeat is not a prerequisite for the binding of the COUP transcription factor. Finally, a 28-mer oligonucleotide which has C to G transversions at positions -81 and -74 did not bind to the COUP transcription factor. This result indicates that the C at position -81 and/or -74 is important and suggested that 28 nucleotides of COUP sequence alone is

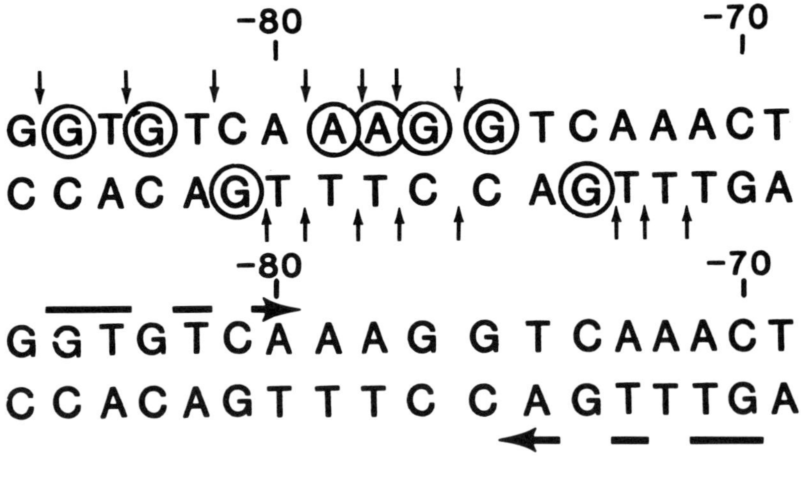

FIGURE 5. Summary of methylation and ethylation interference data. The circle denotes the purine residue which is important for bind and the arrows denotes the phosphate contacts for the COUP transcription factor. The bottom panel indicates a two-fold rotational symmetry.

sufficient for the protein to recognize and bind the ovalbumin promoter.

Methylation and Ethylation Assays

The COUP transcription factor contacting sites were probed in greater detail by methylation and ethylation interference assays. DNA fragments containing COUP sequences were treated with dimethyl sulfate, which methylates N7 of guanine in the major groove and N3 of adenine in the minor groove, or with ethyl nitrosourea, which ethylates phosphates of the DNA backbone. Results are summarized in Fig. 5. These data again indicated that the sequences including the upstream GTCAAA is critical for binding. The contact sites surround the GTCAAAG sequence on both the coding and the non-coding strands, particularly the G residues at positions -85 and -83 of the coding strand and the T residues at positions of -80, -81, -82 of the non-coding strand.

DISCUSSION

In this paper we describe purification of the COUP transcription factor using a specific DNA affinity column containing concatemerized binding sites. This convenient and rapid method enabled us to purify the COUP transcription factor ~700-fold in one step. Thus, the COUP transcription factor has been purified over 200,000-fold from crude extracts of HeLa cells. It should be pointed out that this number is a rough estimation, since at earlier stages of purification, a large amount of pBR322 competitor DNA was used which may result in an underestimation of the COUP factor in those fractions.

The silver-stained SDS gel analysis illustrated that the major COUP binders are 53 kD and 43-48 kD polypeptides. After elution from gel slices and renaturation, all of the protein species still possessed some specific binding. Furthermore, the 43-45 kD proteins could enhance transcription activity. The molecular size of the active component reported here is consistent with our earlier report in which the predominant monomeric protein species was bound and UV-crosslinked to an end-labelled oligonucleotide and estimated to be about 45-50 kD in size (Sagami _et al_., 1986). In addition, S300 column and glycerol gradient centrifugation analyses suggested also that the native protein has a molecular weight of ~90,000 to 100,000. We suggested that the native functional molecular species may be a dimer consisting of two 43-53 kD subunits. Recently, we analyzed the purified COUP peptides by two-dimensional gel electrophoresis (unpublished data). All polypeptides of 43-53 kD displayed similar silver-stain color characteristics and had an identical isoelectric point, suggesting that these polypeptides are structurally related as a result of proteolysis. Finally, we should point out that we do not ignore the possibility that these multiple polypeptides represent either a family of distinct but closely related proteins capable of binding to the same sequences or perhaps the same peptides but having separate forms as a result of post-translational modifications.

The alkylation interference data revealed the exact-contact points on the ovalbumin promoter. It appears that the COUP factor binds to the major groove of the DNA helix in one full turn. Thus, it wraps around the DNA helix instead of binding to one face of the DNA helix. Since the contact sites do disply a two-fold rotational symmetry, one can view the COUP factor as a cresant shape molecule bending in the middle to accommodate for symmetric binding, one half binds to half turn of helix and the other half binds to the second half-turn of helix. It is feasible to view the COUP transcription factor as a dimer having one subunit bind to half-turn of the helix in the major groove and another subunit binds to half-turn of the helix.

As indicated in Figure 1, at least four cell fractions, DE175, DE500, COUP and S300-II are required for faithful and efficient transcription of the ovalbumin gene in an _in vitro_ reconstituted system. Since COUP transcription factor is a sequence specific DNA binding protein and S300-II is not required for transcription of some polymerase II transcribed genes (Tsai _et al._, 1987), it is logical to assume that the DE175 and DE500 fractions provide at least some of the common transcription factors required for formation of the initiation complex. These common factors, TFIID (TATA binding protein), TFIIB, TFIIE and RNA polymerase II are important for initiation of transcription of polymerase II transcribed genes as documented by Roeder and his co-workers

(Hawley and Roeder, 1985; Sawadogo and Roeder, 1985; Reinberg and Roeder, 1987; Reinberg et al., 1987).

We would suggest that the production of COUP factor at a particular developmental stage of differentiation could activate a set of cellular genes, perhaps in a differential quantitative manner related to the relative affinity of COUP factor for each of their slightly divergent promoter sequences. Following such binding, the COUP transcription factor may serve as a nucleation site for other factors, or alternatively to induce conformational changes in the DNA and thus allow other transcription factors to bind nearby. Subsequently, another factor S300-II, RNA polymerase II and other transcription factors may interact in a manner which promotes initiation of transcription of the ovalbumin gene (Tsai et al., 1987).

ACKNOWLEDGMENTS

We thank Peter Hine, Geoffrey Long and JoAnn Bishop for their excellent technical assistance. In addition, we thank the members of the Tsai lab for helpful discussion and Cheryl McCarthy in the preparation of this manuscript. This work was supported by grants HD08188 (BWO) and HD-17379 (MJT) from the National Institutes of Health.

REFERENCES

Arndt-Jovin DJ, Jovin TM, Bahr W, Frischauf AM, Marouardt M. (1975). Eur. J. Biochem. 54:411-418.

Briggs MR, Kadonaga JT, Bell SP, Tjian R (1986). Science 234:47-52.

Carthew RW, Chodosh LA, Sharp PA (1985). Cell 43:439-448.

Chodosh LA, Carthew RW, Sharp PA (1986). Mol. Cell. Biol. 6:4723-4733.

Cohen RB, Sheffney M, Kim CG (1986). Mol. Cell. Biol. 6:821-832.

Dynan WS, Sazer S, Tjian R, Schimke RT (1986). Nature 319:246-248.

Elbrecht A, Tsai SY, Tsai MJ, O'Malley BW (1985). DNA 4:233-240.

Fried M, Crothers DM (1981). Nucleic Acids Res. 9:6505-6525.

Graves BJ, Johnson PF, McKnight SL (1986). Cell 44:565-576.

Hawley DK, Roeder RG (1985). J. Biol. Chem. 260:8163-8172.

Hager DA, Burgess RR (1980). Anal. Biochem. 109:76-86.

Hochschild A, Ptashne M (1986). Cell 44:925-933.

Jones KA, Tjian R (1985). Nature (London) 317:179-182.

Jones KA, Kadonaga JT, Rosenfeld PJ, Kelley TJ, Tjian R (1987). Cell 48:79-89.

Jones KA, Yamamoto KR, Tjian R. (1985). Cell 42:559-572.

Kadonaga JT, Tjian R (1986). Proc. Natl. Acad. Sci. USA 83:5889-5893.

Knoll BJ, Zarucki-Schulz T, Dean DC, O'Malley BW (1983). Nucleic Acids Res. 11:6733-6754.

Parker CS, Topol J. (1984). Cell 37:273-283.

Pastorcic M, Wang H, Elbrecht A, Tsai SY, Tsai MJ, O'Malley BW (1986). Mol. Cell. Biol. 6:2784-2791.

Reinberg D, Roeder RG (1987). J. Biol. Chem. 262:3310-3321.

Reinberg D, Horikoshi M, Roeder RG (1987). J. Biol. Chem. 262:3322-3330.

Sagami I, Tsai SY, Wang H, Tsai MJ, O'Malley BW (1986). Mol. Cell Biol. 6:4259-4267.

Sakonju S, Brown DD (1982). Cell 31:395-405.

Sawadogo M, Roeder RG (1985a). Cell 43:165-175.

Sawadogo M, Roeder RG (1985b). Proc. Nat. Acad. Sci. 82:4394-4398.

Topel SD, Ruden M, and Parker CS (1985). Cell 42:527-537.

Tsai SY, Tsai MJ, Kops LE, Minghetti PP, O'Malley BW (1981). J. Biol. Chem. 255:13055-13059.

Tsai SY, Sagami I, Wang H, Tsai MJ, O'Malley BW (1987). Cell In press.

Wiederrecht G, Shuey DJ, Kibble WK, Parker CS (1987) Cell 48:507-515.

Cellular Factors in Development and Differentiation:
Embryos, Teratocarcinomas, and Differentiated Tissues, pages 53–76

Mapping cis-DNA Regulatory Elements in Seminal Vesicle Secretion Genes, Prostate C3(1) Gene, and the Mouse Mammary Tumor Virus-Long Terminal Repeat DNA

S.E. Harris*, M.G. Parker†, P. Webb†, M. Needham†, Roger White†, J. Ham† and M.A. Harris*, *W. Alton Jones Cell Science Center, Inc., Lake Placid, New York 12946 and †Imperial Cancer Research Fund, London WC2A 3PX, England

INTRODUCTION

Steroid hormone regulated genes have provided a wealth of information on DNA elements which interact with steroid receptors and possibly modulate their rate of transcription (6,7,15,16). The prostate C3 (1) gene and the rat seminal vesicle secretion IV and VI are both regulated by androgens in the sense that their respective steady-state mRNA levels are rapidly induced by testosterone (4,5). The relative roles for transcriptional control and translational controls, such as translational efficiency and mRNA degradation are presently not well defined. By nuclear run off experiments, testosterone increases about 5x the transcription rate of the rat prostate C3 (1) gene (4) and testosterone causes the induction of specific DNase I hypersensitive sites in the rat seminal vesicle secretion gene IV in seminal vesicles nuclei and not in liver nuclei (2). These experiments suggest that the transcription machinery of the rat prostate C3 gene and SVS IV gene are altered to more efficiently transcribe these two genes in the presence of testosterone.

As has been applied to other systems, a search for androgen response elements in the putative control regions of the prostate C3 gene and the SVS IV gene was undertaken. The methodology was basically to prepare plasmids with 5' flanking regions of C3 and SVS IV ligated to a reporter gene such as chloramphenicol acetyl transferase (CAT). Some constructions were made with heterologous promoters, such as thymidine kinase (tk). These plasmid constructions were then transiently transfected into a variety of cell lines which are known to have functional androgen receptor (6,8).

The mouse mammary tumor virus long terminal repeat is known to have a hormone response element (HRE) which interacts with glucocorticoid and progesterone receptor complexes (7). By transient gene transfer experiments, the HRE was shown to be functional *in vivo* (10,11,12,13,6). It has also been shown that this MMTV HRE may contain an androgen response element (ARE) when transfected into androgen receptor (AR) containing cells (13). The use of MMTV-LTR thus provides an ideal model system for mapping and quantitating the strength of various putative ARE's in a variety of cell lines, as well as other modulating cis/trans interactions. Actual identity of the DNA sequence of an ARE will most likely require pure AR/DNA footprinting. The *in vitro* transfection experiments will give us an idea which DNA areas to investigate. We will present data on mapping ARE's in MMTV, rat prostate C3 gene and seminal vesicle secretion (SVS) genes using DDT1 cells derived from a hamster ductus deferens tumor; T-47D^{11}, a human breast cancer cell line; and ZR-75, another human breast cancer cell line (4,5,6). All of these cell lines have functional androgen receptor by a variety of criteria. By sequence comparisons with the HRE in MMTV LTR, we suggest several candidates for the ARE sequences.

We will also present data on transfection of SVS IV 5'-flanking-CAT construction into primary rat seminal vesicle epithelial (rSVE) cell cultures where potential tissue specific cis DNA element may be identified. By DNA sequence comparison of a new seminal vesicle secretion gene, SVS II, with 5'-flanking regions of SVS IV, V, and IV, we propose two DNA sequences as potential tissue-specific cis DNA elements.

MATERIALS AND METHODS

The basic CAT vector for the work with T-47D and ZR-75 was a derivative of PSV/CAT (14) in which the SV40 splice and polyadenelation signals and the CAT gene were placed into the polylinker of pSP65 (3,4). Various upstream sequences were inserted in the polylinker or Hind III site to provide enhancer and/or promoter function. For details of construction, see references 3 and 4. The -1225 to +268 of MMTV-LTR was cloned into PSVØ CAT and used to produce a -201, -501, and -107 deletions of the MMTV-LTR -1225 to +268 fragment. A thymidine kinase promoter was placed in PSVØ CAT at the Pstl site as a 250 bp Pstl fragment, upstream of CAT. This plasmid, PTK-CAT was then linearized with XbaI and a XbaI linkered Hind III fragment from -428 to -69 of the MMTV-LTR was then placed in front of the tk promoter.

The basic CAT vector used with the seminal vesicle genes and DDT1 cells was PSVØ CAT, as described (3,5). 5'-Flanking sequences of SVS IV (-523 to -10) were cloned into the Hind III site of PSVØ CAT and deletions were made using Bal 31 nuclease after Stu I digestions. We also prepared unidirectional ExoIII/mung deletions from -523 towards -10 and cloned these fragments into pSVØ CAT.

The upstream region of SVS VI (-348 to +216) was cloned into PSVØ CAT. PSV 2-CAT with the SV40 enhancer/ promoter 5' of the CAT gene was used as a positive control. Details of these plasmids can be found in reference 5.

Transfection Procedures

T-47D and ZR-75 human breast cancer cell lines were grown in DMEM (Dulbecco's modified Eagle's media) and RPMI, respectively in the presence of 10% FCS (fetal calf serum). Cells used for monitoring the effect of steroids on promoter/enhancer function were grown in the presence of 2% charcoal-stripped FCS. Cells were transfected using the $CaPO^4$ precipitation method with approximately 10^6 cells with 10 µg of plasmid for 6 hrs, followed by a 30 sec 25% glycerol shock (4,14). Steroids were added to the media where appropriate at the following concentrations: 1 to 2 x 10^{-8}M triamcinolone acetonide or dexamethasone; 2 to 3 x 10^{-8}M testosterone; 10^{-6}M progesterone, and 3x10^{-8}M estradiol. After 48 hr the cells were collected and assayed for CAT activity (14).

Plasmids transfections into DDT1 cells were performed using the DEAE-dextran method (5,14). DDT1 cells were initially plated on 15 cm plastic tissue culture dishes (Falcon) and grown in 0.1% fetal calf serum (Gibco) with a media of 50% (DMEM) and 50% F-12 Ham's media, supplemented with insulin (I) (5 µg/ml), transferrin (Ts) (10 µg/ml) and selenium (S) (1x10^{-9}M). This media without the serum is referred to as DFITS media.

The DDT1 cell were grown to about 60% confluence. In an experiment with 20 CAT assays, we would have 20-22, 15 cm dishes of DDT1 cells.

Just before (1-2 hrs.) addition of DNA, the media was changed to 10 ml of DF media (without I, Ts, and S). To this 10 mls of DF media plus cells was added the following mixture in 150λ: 10 μg (10λ) of supercoiled (>80%) plasmid DNA in 100 λ of TBS (Tris 25 mM, pH 7.4, 137 mM NaCl, 5 mM KCl, 0.7 mM $CaCl_2$, 0.5 mM, $MgCl_2$, 0.6 mM Na_2 HPO_4), and 40λ of 100 mg/ml DEAE-dextran (Pharmacia) in TBS. The protocol is essentially as described (14).

The plates were mixed and incubated for 2.5 hrs. The media and DNA-DEAE dextran mix was then removed and 4 mls of 15% glycerol in DF was added to each dish, mixed by rocking and allowed to remain on the cells for 120 seconds. The 15% glycerol shock solution was then immediately diluted with 10 ml of DFITS media +5% FCS and then removed. The cells were washed 1X with 15 ml of DF media. 25 ml of DFITS media +5% FCS was then added. After 24 hrs the media was removed, washed 1X with DF and then, 25 ml of DFITS was added, plus or minus steroid hormones. In plates with testosterone, the concentration was 2 x 10^{-8}M. In plates with glucocorticoids, the triamcinolone acetonide concentration was 2X 10^{-8}M. When both steroid hormones were added together, they were at 2X 10^{-8}M each. The DDT1 cells + hormones in serum-free DFITS was then incubated for 72 hrs.

The cells were collected and the cell pellet was then frozen.

Cat Assay

The assay for chloramphenicol acetyl transferase was as described (3,14).

RESULTS

1. Comparison of MMTV promotes activity with androgen, glucocorticoid, progesterone and estrogen in ZR-75 and T-47D human breast cancer cells

The full length -1225 to +268 region of MMTV-LTR was cloned into pSVØ CAT and transiently transfected into these two cell lines and then treated with various steroid hormones. Table I shows the results. The induction ratio (CAT activity in cells treated with steroid/CAT activity of cells not treated with steroid) is compared for the two cells with the different steroids. No estrogen response with MMTV-CAT was seen with either cell line although both cell lines have estrogen receptor. However, (Table 1) there was a 50-150 X increase in CAT activity in extracts from cells treated with

Table 1.

Steroid	Induction Ratio	
	ZR-75	T-47D
Testosterone	85	52
Dexamethasone	48	2
Progesterone	60	145
Estrogen	1	1

The induction ratio represents the ratio of CAT activity obtained with the full-length LTR as promoter in the presence and absence of steroid.

either T (testosterone), G (dexamethasone), or P (progesterone), except the T-47D cells. There appears to be a correlation between AR (androgen receptor) concentration and CAT induction ratio between ZR-75 and T-47D. Table 2 presents the concentration of various steroid receptors in cells used in this study.

The basic conclusion, however, is that the MMTV-LTR (-1225 to +228) has GRE, PRE and AREs (RE = response element). Further mapping is presented below and in reference 4. This study demonstrates that both the ZR-75 and T-47 D cell lines are excellent model systems for mapping AREs, PREs, and GREs and possible overlapping DNA sequences within these hormone response elements.

2. Mapping the androgen-, progesterone-, and glucocorticoid response regions in the MMTV-LTR using the ZR-75 breast cancer cell line

The full length -1225 to +268 MMTV-LTR, the -501 to +268, the -201 to +268, and the -107 to +268, were cloned into pCAT plasmid. These constructions were then transfected into ZR-75 cells and then tested with and without progesterone, glucocorticoids, and androgens. The results are presented in Figure 1, expressed as % CAT activity where 100% CAT activity is for the -1225 to +268 MMTV-LTR-CAT construction. It is clear that there is at least one ARE, GRE, and PRE between -201 and -107 in the MMTV-LTR, since deletion of this region results in loss of steroid response for all three steroids. This confirms previous results (7) and demonstrates that the ZR-75 cells are again a very

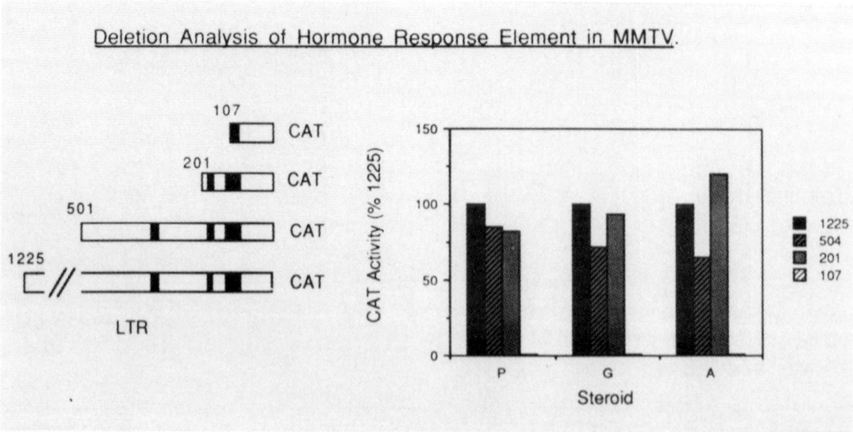

Figure 1. Steroid Hormone Response Elements (HRE) in MMTV-LTR. Deletion of the -1225 to +268 regions of the MMTV-LTR ligated in front of the CAT gene were transfected into ZR-75 cells and tested with and without testosterone (A), glucocorticoids (G), or progesterone (P). Relative CAT activity in cell extracts is presented where 100% is for the -1225 to +268-CAT plasmid.

useful cell line for further mapping androgen response elements (AREs).

3. Steroid hormone regulated enhancer function of the -428 to -69 region of MMTV-LTR

The -428 to -69 Hind III fragment was cloned upstream of the tk promoter in the plasmid pTK-CAT. It was cloned in both directions to test for steroid dependent enhancer function by transfection into ZR-75 cells and T47D cells. The results are presented in Figure 2. This -428 to -69 HRE containing region of MMTV-LTR could serve as a testosterone, glucocorticoid, and progesterone dependent enhancer element, as noted by the induction ratios in Figure 2. Likewise, this HRE (-428 to -69)-TK-CAT, when transfected into T47-D cells could serve as a steroid dependent enhancer element for testosterone and progesterone, but was inactive with glucocorticoids.

Table 2.

Steroid Hormone Receptor
Concentration in Various Cell Lines

	Androgen Receptor	Glucocorticoid Receptor	Progesterone Receptor	Estrogen Receptor
		fmoles/mg protein		
1. ZR-75 (human breast cancer)	590	140	260	55
2. T47-D#1 (human breast cancer)	230	45	510	90
3. DDT1-MF2 (hamster ductus deferens tumor)	400	1200	-	-

This oligonucleotide, -192 to -165 of the MMTV-LTR HRE, could serve as a weak testosterone dependent enhancer element (~ 2X induction ratio). The rest of the androgen response element(s) may reside elsewhere. The -192 to -165 oligonucleotide could serve as a fairly strong glucocorticoid or progesterone dependent enhancer element. However, the -192 to -165 oligonucleotide was not as nearly as strong a steroid-dependent enhancer element as the whole -428 to -69 HRE (induction ratio of 16 to 80) for any of the three hormones. From the data of Table 3, there is the suggestion that part of an ARE may reside in the -192 to -165 and that a major GRE is also in this sequence. Only part of the progesterone stimulated CAT activity can be accounted for by this -192 to -165 region of the MMTV-LTR HRE.

5. <u>Search for AREs in the prostate C3(1) gene</u>

Several regions of the C3(1) upstream sequences were cloned into pSVØ CAT equivalent or pTK-CAT. The -750 to -110 region, the -750 to +20 region, and the -7000 to +3000 region were tested and found to have little or weak androgen responsiveness when transfected into two prostate cell lines which give weak androgen responsiveness with the control HRE-CAT (-428 to -69) construction. Further constructions and transfection into other cell lines may help elucidate if functional AREs or androgen-dependent cis DNA elements are present in the C3(1) gene (4).

6. <u>Transfection of hamster DDT1 cells with rat seminal vesicle secretion gene (SVS) IV and VI 5'-flanking region -CAT Plasmids and the effect of testosterone and glucocorticoids</u>

A -523 to -10 region of SVS IV gene and a -348 to +216 region of SVS VI gene were cloned into the Hind III site of pSVØ CAT and referred to as p24 CAT and p14 CAT, respectively. The MTV-LTR-(-1225/+268) CAT plasmid was previously described. A BAL 31 nuclease deletion of p24 CAT (Δ from -360 to -90) was also prepared by StuI (-172) digestion, BAL 31 digestion, and religation. Several exonuclease III/mung bean unidirectional deletions, 5' → 3' of the -523 to -10 region of SVS IV were prepared and then cloned into the Hind III site of pSVØ CAT.

The Syrian hamster ductus deferens tumor cell line, DDT_1-MF_2 contain high levels of androgen receptor and glucocorticoid receptor (Table 1). The DDT_1 cells are growth-stimulated by testosterone at the 10^{-9} to 10^{-7} M) and glucocorticoids can block this androgen induced stimulation of growth (8,9). Using the plasmid constructions, p 24-CAT (SVS IV - 523 to -10) and p14 CAT (SVS VI -348 to +216) and a BAL 31 deletions (pΔ 20-2) of the SVS IV 5'-flanking region and MTV-LTR-CAT, we performed transient transfection experiments into these DDT_1-MF_2 cells in the presence or absence of testosterone (T @ 2 x 10^{-8} M) and in the presence of T and the glucocorticoid, triamcinalone acetonide, at 2 x 10^{-8} M (TAA). These results are presented in Figure 3.

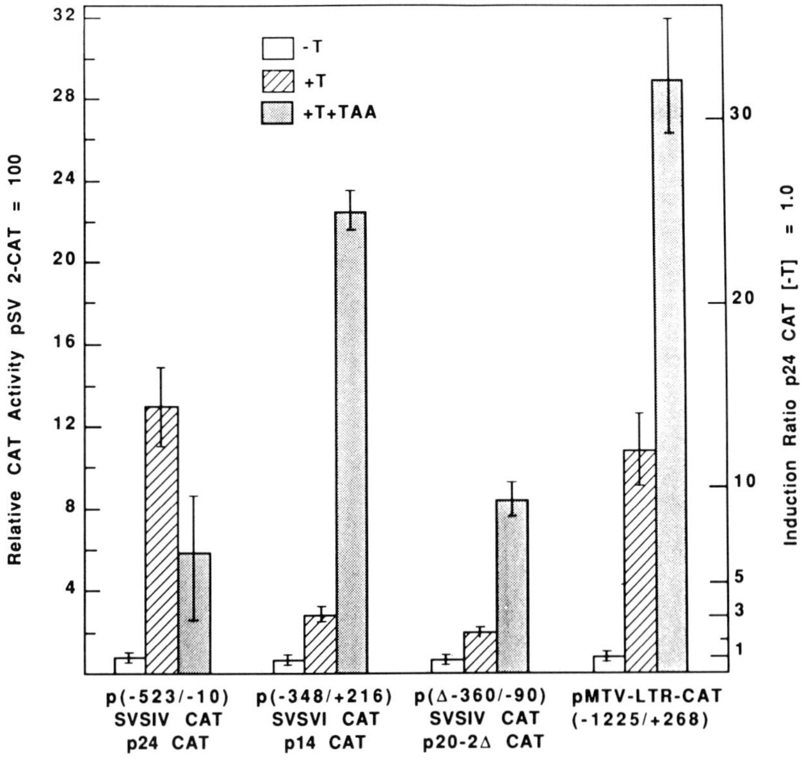

Figure 3. Seminal vesicle gene/CAT constructions and MTV-LTR/CAT were transfected into DDT1-MF2 cells cultured in the presence and absence of testosterone(T) and testosterone plus the glucocorticoid, triamcinalone acetonide (TAA). The data are expressed relative to pSV2-CAT which has the SV40 promoter/enhancer region.

Briefly, pSV2-CAT (SV40 enhancer and promoter ligated to CAT) was added to ~10^6 DDT_1-MF_2 cells (~60% confluent) previously grown in DMEM: F12 1:1, insulin, transferrin and selenium (DFITS Media) plus 0.1% serum using the DNA-DEAE-dextran method (25). On day 2, steroids were added and serum removed for 72 hr. DDT_1 cell extracts had easily measurable CAT activity with pSV2 CAT. pSVØ CAT had approximately 50 to 100x less CAT activity. The SVS-CAT construction, p24 CAT (-523 to -10 SVS IV-CAT), in these same experiments showed a 14x stimulation in CAT activity when the transfected DDT_1 cells were grown in the serum-free media (DFITS), in the presence of 2 x 10^{-8} M T. There was a decrease in CAT activity when both T and TAA was present in the media. This suggests negative effect of glucocorticoids on SVS IV expression.

Cell extracts from transfections with the BAL 31 deletion, p 20-2/ SVS IV (Δ -364 to -90) gave no increase in CAT activity when T was present, but did give a 7x stimulation when both hormones, T and TAA were present in the media (Figure 3). This suggests the negative glucocorticoid responsive area, either a GRE or some other indirect effect is in between -90 and -360 in the 5'-flanking region of SVS IV. The SVS VI gene, p14 CAT (-348 to +216 SVS VI ligated to CAT) gave a 2-3x increase in CAT activity with T, indicating a weaker ARE or possibly a compounding negative cis element. Surprisingly, in the presence of both T and TAA, p14 CAT gave a 25x stimulation in CAT activity. This indicated there may be a strong GRE in the SVS VI gene, or TAA is modulating other factors involved in SVS VI expression. By sequence homology to the consensus sequence for a GRE, TGTPyCT, we found two tandem copies in an inverted orientation in the 1st intron of SVS VI gene at +135.

The control plasmid, pMMTV-LTR-(-1225/+268)-CAT was transfected into the DDT1-MF2 cells, as described. Cell extracts from DDT1-MF2 cells treated with testosterone (T) or testosterone (T) plus glucocorticoids (TAA) were assayed for CAT activity and showed a 11x stimulation with T alone and a 32x stimulation with both T and TAA (Figure 3). This experiment demonstrates the DDT1 cells have functional androgen- and glucocorticoid receptors which can interact with the HRE element in the MMTV-LTR. It suggests that the two receptors may be functioning through different cis DNA elements if we assume the ARE sites are fully saturated when glucocorticoid is added to the system. Competition assays with various double-strand oligonucleotides to regions in the HRE, as well as footprinting with pure AR and GR and this HRE will help resolve these matters.

In summary, this DDT_1-MF_2 cell system seems ideal for quickly mapping ARE's and GREs in these short 5'-flanking and exon/intron regions of the SVS IV and VI genes. Also, other trans-acting factors which are effected by T and/or TAA may be elucidated. At this point we do not known the sequence for an ARE so our conclusions are tentative.

7. Transfection of pSVS-CAT plasmids into primary rat seminal vesicle epithelial cell cultures

With the increased interest and potential importance of tissue-specific trans-acting factor(s), as well as the steroid receptor as an inducible transacting enhancer protein (15,16,17), we turned to developing *in vitro* tissue culture conditions to get rat seminal vesicle epithelial (SVE) cells to: 1) growth in sufficient quantity to be able to do transient transfection studies and 2) have these SVE cells respond to testosterone *in vitro* by increased production and secretion of some of the SVS proteins (i.e., see ref. 18,19). This is central to many of the questions we are now asking, concerning the mechanism of the tissue-specific production and secretion of the SVS IV, VI and II proteins, as well as their androgen regulation.

A comparison of growth factor requirements for prostate (PE) and SVE cells grown in WAJC 404 media on *plastic* is presented in Figure 4. These SVE and PE cells were from 35 day-old rats (post-puberty). We found that SVE cells from this stage animal required acidic ECGF or FGF (HBGF-I) for optimal growth (HBGF-I = heparin-binding growth factor), as did the PE cells (21,22).

Taking the information available on growth and differentiation of breast epithelial cells (see eg. ref. 18,19), we found that early or pre-pubertial animals (25-30 days), where the rat seminal vesicle (both lobes) only weighs approximately 40 mg and there is no apparent secretion at this stage, when placed in primary culture in WAJC 404

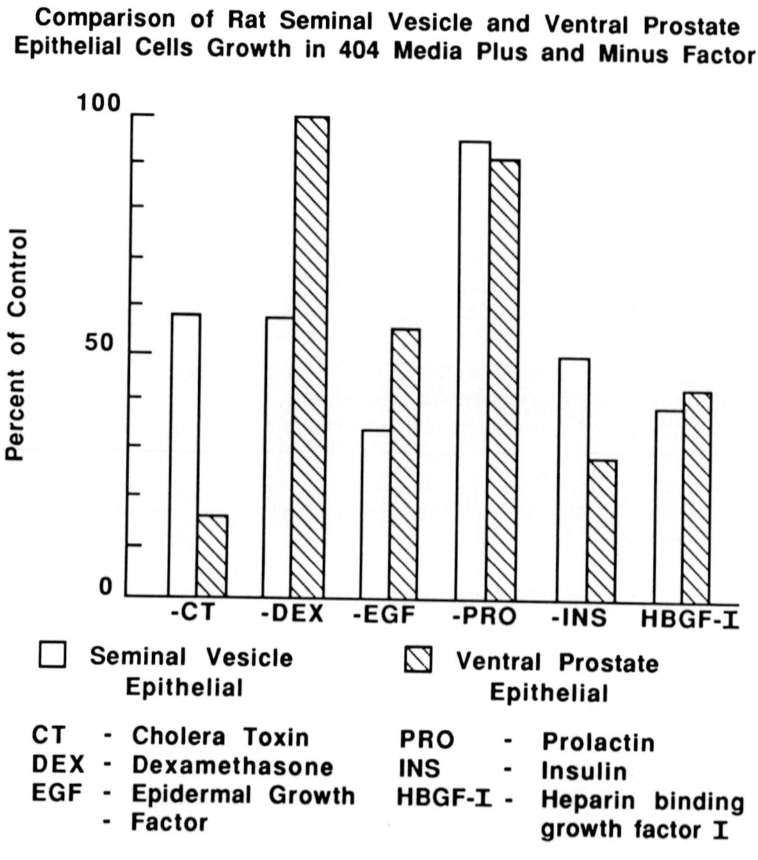

Figure 4. Rat seminal vesicle and ventral prostate tissue were treated with collagenase and prepared for primary culture in WAJC 404 media on plastic (21). The effect of various growth factors on growth was tested for both the prostate epithelial cells and the seminal vesicle epithelial cells, as described (21).

media, gave the largest number of SVE cells per animal after 5-7 days in culture. We could get approximately 0.5 x 10^6 cells after 5 days of culture of SVE cells derived from 20 animals. Initial experiments of ^{35}S-methionine labeling (4-6 hr) of these cultures at 7-10 days (grown from the first day with or without 10^{-7} testosterone), indicated they <u>were not</u> selectively synthesizing detectable SVS proteins or any other major low molecular weight proteins in the presence of testosterone. The overall rate of incorporation of ^{35}S-methionine into protein was the same whether testosterone was present or not.

a. <u>Transient transfections of pSVS-CAT constructions into SVE cells</u>

However, when we transfected a CAT construction plasmid consisting of -523 to -10 bp of the 5'-flanking region of SVS IV gene ligated into the Hind III site of pSVØ CAT (14), p24 CAT, into these cultures on plastic, we found that cell extracts from SVE cells grown with 5 x 10^{-8} M testosterone (T) (from Day 1) contained about 10-15x more chloramphenicol acetyl transferase (CAT) activity (Figure 5). A BAL 31 deletion from -362 to -37 of the p24 CAT plasmid was constructed and called pΔ10-1. When this pΔ10-1 was transfected into SVE cells ± T, on plastic and SVE cell extracts were assayed for CAT, there was no change or a slight decrease in CAT activity when the extract from the cells grown in the presence of T was assayed for CAT. This indicated that between -362 and -37 of the SVS IV gene 5'-flanking region, there were cis element(s) involved in androgen stimulation of this SVS IV gene. The positive control plasmid, pSV2 CAT, with SV40 enhancer and promoter was approximately 50x that of p24 CAT minus testosterone in this system. These results are presented in Figure 5.

We also cloned the -348 to +216 region of SVS VI gene into pSVØCAT and tested it in the SVE cell system on plastic. It gave a 3x increase in CAT activity when T was present. The basal or -T CAT activity for this SVS VI construction was 3-4x higher than p24 CAT activity (-T) (Figure 5). This may be due to the dexamethasone (10^{-6} M) in the WAJC 404 culture media and possible glucocorticoid effects on the SVS VI gene, as was observed with the DDT1 system.

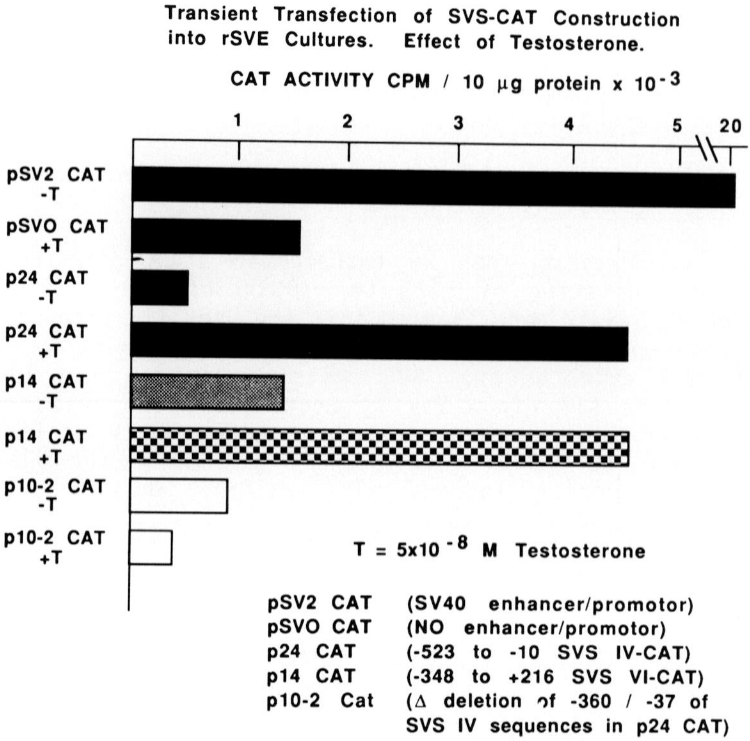

Figure 5. Various CAT plasmids, as described above, were transfected into primary rat seminal vesicle epithelial cell cultures grown in the presence and absence of testosterone(T) on plastic in WAJC 404 media. CAT activity was measured in cell extracts, as described in Methods.

b. <u>Growth and "differentiation" of SVE cells in WAJC 404 media on an extracellular matrix</u>

When early pubertial (28-35 day) rats are used for seminal vesicle tissue for primary culture, and the cells are placed in WAJC 404 media plus EGF, CT (cholera toxin), DEX, Insulin and ± T, on a <u>gel</u> of "Matrigel", the SVE cells aggregate and grow as 3-dimensional epitheloids. However, the SVE cells aggregate, attach, and grow as thick sheets of epithelial cells when placed on a <u>film</u> of Matrigel in the WAJC 404 media.

Several interesting observations have been made with this "young" SVE cell culture system on Matrigel films with WAJC 404 + Factors ± Testosterone. One, the cultures continue to grow and remain fairly stable for up to 2 weeks. Two, when cultures are grown for 7-14 days ± T and then "pulsed" with ^{35}S-methionine in WAJC 404-methionine media for 4-6 hr, the cultures of SVE grown with T, synthesize and secrete to a limited extent radioactive proteins in the low molecular weight region of a Laemmli-SDS gel, consistent with T-induced secretory proteins of the seminal vesicle (e.g., SVS IV, V, II and VI). We are now extending these experiments using antibody to synthetic peptides to SVS IV, VI and II and <u>in situ</u> immunofluorescence and Western blots.

Another phenomena with the SVE cultures grown on Matrigel films is that the cells grown in the presence of testosterone (10^{-7} M) for 9 days are capable of much higher rates of ^{35}S-Methionine incorporation into both cellular protein (3-4x increase) and into protein apparently secreted into the media (10x increase) (Table 4).

By phase contrast microscopy, the SVE cell appears to be denser and full of ribosomes (Figure 6). This has not been confirmed directly. The yield of RNA from the approximate same number of SVE cells grown in testosterone is 2-3x that obtained from the SVE cells grown without testosterone. These preliminary data support the hypothesis that our culture conditions with young SVE cells on a Matrigel film can respond to testosterone in a way somewhat similar to the <u>in vivo</u> situation (23). Further optimization of this rSVE culture system is continually being developed.

Table 4.

INCORPORATION OF ^{35}S-METHIONINE INTO PROTEIN OF RAT PRIMARY SEMINAL VESICLE EPITHELIAL CELL (SVE) CULTURES GROWTH ON EXTRACELLULAR MATRIX*

Hormone Treatment (9 days) Testosterone 10^{-7} M	C.P.M. x $10^{-6}/10^{4}$ cells or "secreted" from 10^{4} cells
1. SVE Cells -	125
2. SVE Cells +	407
3. "Secreted" Protein -	3.9
4. "Secreted" Protein +	42.3

*Rat SV primary cultures grown on extracellular matrix in serum-free WAJ404 media, EGF, cholera toxin, and insulin for 9 days either with or without 10^{-7} M testosterone. The cultures were then labeled with 750 μCi of ^{35}S-methionine (1142 Ci/mmole) for 4 hrs in 1.0 ml of WAJ404 media made up minus methionine. The supernatant was precipitated (50 μg/ml BSA carrier) with cold acetone and washed twice, dried and dissolved in Laemlli buffer. The cell pellet was washed twice with PBS and lyzed in 0.15 M NaCl, 10 mM Tris-HCl, 5 mM EDTA. Equal quantity of C.P.M. was then applied to a SDS-Laemlli gel for electrophoresis.

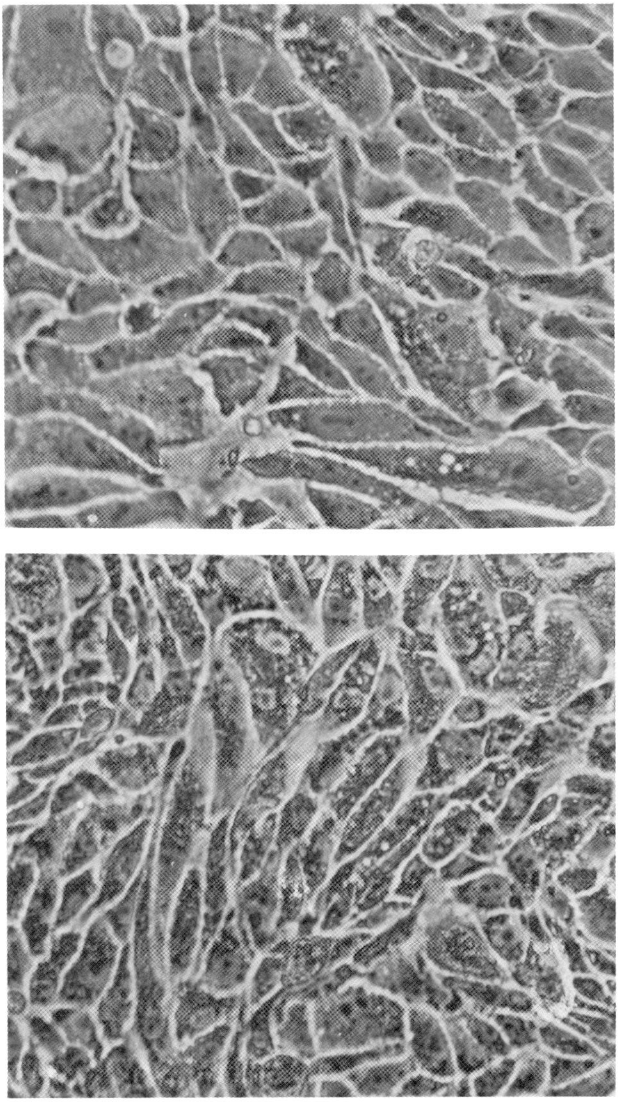

Figure 6. Phase contrast photomicrograph of rat primary seminal vesicle epithelial (rSVE) cells grown on "Matrigel"-extracellular matrix in the presence (Bottom) and absence (TOP) of 5×10^{-8} M testosterone in WAJC 404 media. Tissue was from "young" 32 day-old animals and the rSVE cells were in culture for 7 days.

8. The SVS II gene

Robert Matusik at the University of Manitoba, Winnepeg, found a cDNA clone in rat dorsal prostate which was capable of measuring the androgen- dependent induction of seminal vesicle secretion mRNA, SVS II mRNA, in both dorsal prostate and seminal vesicle (24). This cross-tissue specificity is extremely interesting. With this cDNA to presumably SVS II (~300 bp), we isolated from a rat Charon 4A (EcoR1) library a genomic clone for the SVS II gene. Within a cloned 14 kb piece of rat DNA, we found a 0.95 Kb Bam fragment that has sequences identical with sequences in the SVS II cDNA.

We purified ~10 mg of SVS II, and in collaboration with John Crabb, protein chemistry lab, his group has sequenced over 270 amino acids (a.a) out of ~400 a.a. of the total 45,000 mol. wt. SVS II protein. We prepared two oligonucleotides to SVS II peptides and showed that within a 7.3 kb EcoR1 of the λ Charon 4A-rat DNA clone, both oligonucleotides specifically hybridized.

We have at this stage subcloned the 7.3 kb EcoR1 fragment into pBlue-Script and analyzed 30 deletion clones, using the pBlue-Script/Exo III/Mung Bean Nuclease system of Stratagene, by DNA sequence analysis.

c. Sequence analysis of SVS II

Over 5.0 kb of DNA sequence has been completed on the SVS II gene in this 7.3 kb EcoR1 fragment. This fragment has 5,000 bp of 5'-flanking region, the 1st exon, the 1st intron, the 2nd exon, and part of the 2nd intron. Approximately 2,000 bp DNA sequence of the 5'-flanking region has been obtained.

The 1st exon contains the leader peptide and 4 amino acids of the secreted protein. This is very similar in structure to the 1st exon of SVS IV, V and VI. The 2nd exon contains all of the coding information for the remaining secreted SVS II protein - 388 amino acids. We are most likely missing part of the 2nd intron and 3rd non-coding exon, if the SVS II gene structure is similar to SVS IV, V and IV. The SVS II protein sequence is unique. No outstanding homologies were found in a comparison with GenBank™ 532 or PIR™ 14 sequences. We noted a series of 13 amino repeats in the structure which accounts for over one-half of the protein. This sequence is probably involved in the transglutamase cross-linking reactions during the SVS I, II and III clotting reaction. The repeat contains 2 to 4 glutamines. Further analysis of this sequence may give us some insight into the mechanism of coagulation and clotting.

The apparent lack of introns in the major coding region is intriguing. At this point we do not have any speculations, except that it is unusual for such a large protein to be coded for mostly by one exon. A general map of SVS II is presented in Figure 7.

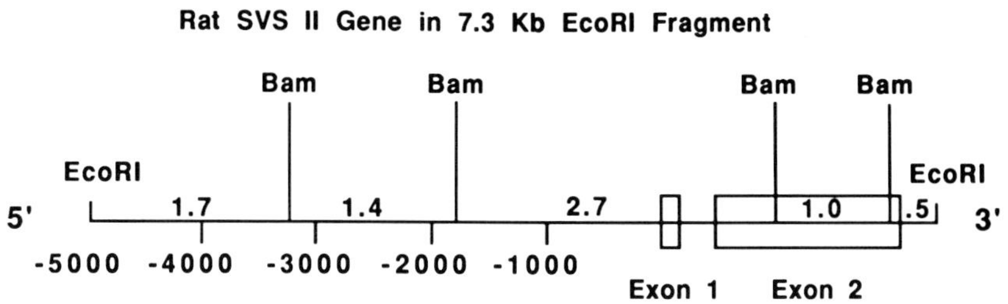

Figure 7. The rat seminal vesicle secretion II gene (SVS II). Schematic drawing of the SVS II gene in a 7.3 kb EcoR1 fragment isolated from a Charon 4A rat genomic library. Approximately 5,000 bp of 5'-flanking region are in this DNA fragment. Boxes indicate exon regions so far identified in the SVE II gene.

9. 5'-Flanking Region Homologies in SVS II, IV, V and VI

When we analyzed the 5'-flanking region of SVS II for homologies to SVS IV, V and VI, we found three very interesting areas: 1) the CATTT homology at -90 area in SVS IV, V and VI is found at -110 in SVS II; 2) there is an inverted repeat at about -140 in the SVS II gene, similar to the ones found in SVS IV, V and VI. In fact, we found to our surprise, the loop area of the stem-loop structure in SVS II has 10/11 bases that are identical to the same loop area in SVS V and VI (A C T G A G C A G A G); 3) in the -200 region of SVS II, there is a 13 bp identical sequence in SVS IV.

Less similar (although present, 6/6 for SVS V; 10/10 for SVS VI) are present in this area of SVS V and VI. This -200 region also has high homology to a GRE/ARE region in MMTV-LTR. The presence of these conserved sequences in SVS II, IV, V and VI has given us confidence that we may be looking at significant areas which will play a cis-DNA role in regulation of this set of seminal vesicle secretion genes. In Table 5 is presented the sequences of these three areas in SVS II, IV, V and VI.

Table 5.

Homologous Regions In the 5'-Flanking Region of Four Rat SVS Genes

A. Region -200 GRE/ARE Homology

SVS II	-226 to -214	G G T T G A T A A A G A T T
SVS IV	-210 to -198	T G T T G A T A A A G A T T
SVS V	-188	T A A A A A G G A A G A T T
SVS VI	-184 to -175	A A A G A T A A A G A T T
HRE-MMTV	-177 to 165	T G T T C T T A A A

B. Region - Inverted Repeat/Homology

SVS II	-139 to -119	T C C C C A A C T G T G C A G A G G G G A
SVS IV	-123 to -102	C C C C C T G C A T A G C C C A G G G G G
SVS V	-120 to -96	C T C C T A C T G A G C A G A G A G A A A G G A G
SVS VI	-119 to -96	C T C C T A C T G A G C A G A G A G A - A G G A G

C. Region - C A T T T Homology

SVS II	-109 to -97	C A T T T A G A G A T A A
SVS IV	-92 to -80	C A T T T A C A G A T A A
SVS V	-90 to -78	C A T T T A C C G A T A A
SVS VI	-90	C A T T T A C T G A T A A

DISCUSSION

With the functional CAT assay of various plasmid/ transient transfections into a variety of cell lines, the region in the HRE of MMTV-LTR was shown to function as an androgen-dependent enhancer. As well, analysis of a synthetic oligonucleotide of -192/-165 of the HRE showed it retained much of the glucocorticoid activity, a good bit of the progesterone activity, and some of the androgen responsiveness. This oligonucleotide may only contain part of an ARE. Once fairly pure androgen receptor is available, footprinting experiments should help resolve these matters.

The prostate C3(1) gene doesn't seem to have any strong androgen responsive areas when transfected into prostate cell lines. Once an idea of what an ARE looks like, regions in the C3(1) gene similar to the ARE can be selectively tested in cells such as DDT1 or ZR-75, where a strong androgen response is seen with the MMTV-LTR HRE.

The rat SVS genes offer a set of androgen-responsive genes in which various regions of the genes have and will be tested in both primary rSVE cell cultures and in androgen receptor containing tumor cells such as ZR-75 and DDT1s. With the DDT1 cells we will be able to map some general cis elements as well as the AREs. We hope to be able to map some tissue-specific cis DNA elements in our primary rat seminal vesicle epithelial cell culture system. By correlating such functional assays with footprinting analysis and gel-shift, DNA-protein interaction studies, we may begin to better understand the role and mechanism of action of androgens on gene expression.

REFERENCES

1. Zhang YL, Parker MG (1985). Mol Endocrinol 43:151.
2. Harris SE, Teng CT (1988). Mol Endocrinol (In press).
3. Gorman CM, Moffat LF, Howard BH (1982). Mol Cell Biol 2:1044.
4. Parker MG, Webb P, Needham M, White R, Ham J (1987). J Cell Biochem 35:285.
5. Harris SE, Schifman AL (1988) J Biol Chem (In press).
6. Scheidereit C, Beoto M (1987). Steroid Receptor Binding to DNA Sequences. In Green B, Leake RE (eds): "Steroid Hormones, A Practical Approach", Washington, D.C.: IRL Press, pp. 179-218.

7. von der Ahe D, Ianich S, Scheidereit C, Renkrowitz R, Schutz G, Beato M (1985). Nature 313:706.
8. Norris JS, Kohler PO (1977). Endocrinology 106:613.
9. Syms AJ, Norris JS, Smith RG (1983) In Vitro 19:929.
10. Zaret KS, Yamamoto K (1984). Cell 38:29.
11. Hollenberg SM, Griquere V, Segui P, Evans RM (1987). Cell 49:39.
12. Vanderbilt JN, Miesfeld R, Maler BA, Yamamoto KR (1987). Mol Endocrinology 1:68.
13. Darbre PD, Moriarty A, Curtis SA, King, RJB (1985). J. Steroid Biochem 23:379.
14. Gorman C (1985). High Efficiency Gene Transfer into Mammalian Cells. In Glover, DM (ed): "DNA Cloning," Washington, D.C.: IRL Press, Vol. II, pp. 143-190.
15. Knoll BJ, Zarvcki-Schulz T, Dean DC, O'Malley BW (1983). Nucleic Acid Res 11:6733.
16. Yamamoto KR (1985). Ann Rev Genet 19:209-252.
17. Walker MD, Edlund T, Boulet AM, Rutter, WJ (1983). Nature 306:557.
18. Yeu-Hua E, Lee WH, Kaetzel CS, Parry G, Bissell MG (1985). Proc Natl Acad Sci USA 82:1419.
19. Chen Li-How, Bissell M (1987). J Biol Chem 262:17247.
20. Hadley MA, SW Byers, Suzrez-Quian CA, Kleinman HK, Dym, M (1985). J Cell Biol 101:1511.
21. McKeehan WL, Adams PS, Rosser MP (1984). Cancer Res 44:1998.
22. Crabb JW et al (1986). Biochemistry 25:4988.
23. Fawell SE , Higgins SJ (1986) Mol Cell Endocr 48:39.
24. Matusik RJ, Kreis C, McNicol P, Sweetland R, Mullin C., Fleming WH, Dodd JG (1986) Biochem Cell Biol 64:601.

Cellular Factors in Development and Differentiation:
Embryos, Teratocarcinomas, and Differentiated Tissues, pages 77–92

Egg-Induced Modifications of the Mouse *Zona Pellucida*

Richard M. Schultz[1], Yoshihiro Endo[2], Peter Mattei[1], Shigeaki Kurasawa[2], and Gregory S. Kopf[2]
Department of Biology[1] and Division of Reproductive Biology and Endocrinology[2], Department of Obstetrics and Gynecology, School of Medicine, University of Pennsylvania, Philadelphia, PA 19104

Mouse eggs, and mammalian eggs in general, are surrounded by an extracellular matrix called the *zona pellucida* (ZP). The initial events of fertilization, which includes species-specific binding of sperm to the *zona* and the acrosome reaction, are initiated on the *zona*. The composition and function of the constituents of the mouse *zona pellucida* and will first be described. Results of some of our recent experiments that concern egg induced modifications of the mouse egg's *zona*, which may be involved in the block to polyspermy, will then be presented.

Composition of the Mouse Egg *Zona Pellucida* and Function of *Zona* Proteins

The mouse egg's *zona pellucida* is composed of three sulfated glycoproteins called ZP1, ZP2, and ZP3 (Bleil and Wassarman, 1980a; Shimizu et al., 1983) and are present in molar ratios of about 1:10:10, respectively. These proteins are synthesized by the oocyte and secreted throughout the period of oocyte growth (Bleil and Wassarman, 1980b). The *zona* proteins are a major biosynthetic product of the growing oocyte, since synthesis of the *zona* proteins constitutes about 15% of total oocyte protein synthesis; their synthesis is essentially undetectable in the fully-grown oocyte (Bleil and Wassarman, 1980b). Recent studies have shown that transcripts for ZP3 accumulate during the period of oocyte growth and comprise about 0.1–0.2% of the total poly (A^+) RNA; upon completion of growth and oocyte maturation the level of ZP3 transcripts falls to about 15% of its maximal level (Philpott et al., 1987).

ZP1 has an apparent molecular weight of about 200,000 daltons (Bleil and Wassarman, 1980a). It is a dimer connected by intermolecular disulfide bonds. A possible biological function of ZP1 is to serve as a cross-linker responsible for maintaining the three dimensional structure of the *zona pellucida* (Greve and Wassarman, 1985). Electron microscopic analysis of *zonae* that have been partially dissolved by either low pH or mild protease treatment indicates that the *zona* is composed of long filaments. ZP1 may connect these filaments, since partial digestion of *zonae* with chymotrypsin degrades only ZP1 and disrupts the filamentous structure (Greve and Wassarman, 1985).

ZP2 isolated from oocytes or unfertilized eggs has an apparent molecular weight of 120,000 daltons, which is observed either under nonreducing or reducing conditions of gel electrophoresis (Bleil and Wassarman, 1980a). Fertilization results in the modification of ZP2 to a form called $ZP2_f$. Under nonreducing conditions, $ZP2_f$ has an apparent molecular weight of 120,000, whereas under reducing conditions it has an apparent molecular weight of 90,000 daltons (Bleil et al., 1981). Thus, the modification probably results from proteolysis in which the two fragments are held together by disulfide bonds. Radiolabeled ZP2 can bind to acrosome reacted sperm, whereas radiolabeled $ZP2_f$ cannot (Bleil and Wassarman, 1986). As wiil be discussed later, ZP2 may mediate sperm binding subsequent to the acrosome reaction.

ZP3 has an apparent molecular weight of 83,000 daltons and accounts for the sperm receptor activity of the *zona pellucida* (Bleil and Wassarman, 1980c). Competitive inhibition of sperm binding to *zona*-intact eggs using purified ZP3 from unfertilized eggs indicates that ZP3 possesses all of the sperm receptor activity present in structurally intact *zonae*. Small glycopeptides derived from ZP3 have also been demonstrated to possess sperm receptor activity (Florman et al., 1984). O-linked carbohydrate moieties are thought to be responsible for the sperm receptor activity of ZP3, since conditions that selectively remove O-linked carbohydrate inactivates sperm receptor activity. Conditions that selectively remove N-linked carbohydrate do not alter sperm receptor activity (Florman and Wassarman, 1985).

ZP3 also possesses the ability to induce the sperm acrosome reaction, and can do so to the same extent as that induced by the ionophore A23187 (Bleil and Wassarman, 1983). The small glycopeptides derived from proteolyzed ZP3, however, cannot induce the acrosome reaction. Although the polypeptide chain may play a direct role in inducing the acrosome reaction, its role may also be to position correctly carbohydrate moieties that are involved in inducing the acrosome reaction; presumably such spatial information is not present in the glycopeptides.

Fertilization of Mouse Eggs and *Zona* Block to Polyspermy

Based on these properties of the mouse *zona* and the observation that only acrosome-intact sperm bind to the *zona*—acrosome-reacted sperm do not bind—fertilization of mouse eggs is comprised of the following sequence of events. Acrosome-intact sperm bind to the *zona pellucida* (Saling et al., 1979; Florman and Storey, 1982) and this species specific binding is mediated by ZP3, which in turn induces the acrosome reaction of these bound sperm. ZP2 mediates the binding of acrosome-reacted sperm (Bleil and Wassarman, 1986), which then penetrate the *zona pellucida* and gain access to the perivitelline space—the space between the *zona pellucida* and the egg's plasma membrane. The acrosome-reacted sperm then bind to and fuse with the egg's plasma membrane.

In response to fertilization, the egg undergoes the "cortical granule reaction" (Szoilsi, 1967; Barros and Yanagamachi, 1971, 1972; Wolf and

Hamada, 1977). It is believed that cortical granules subjacent to the plasma membrane fuse with the plasma membrane and release into the perivitelline space enzymes that convert ZP2 to $ZP2_f$ and modify ZP3 such that it no longer possesses either sperm receptor activity or the ability to induce an acrosome reaction. Thus, acrosome-intact sperm can no longer bind to the *zona* and acrosome-reacted sperm bound to the *zona* can no longer interact with and penetrate the *zona*, since these sperm do not interact with $ZP2_f$ (Bleil and Wassarman, 1986). This series of events is thought to comprise the *zona* reaction or the *zona* block to polyspermy. With time, a membrane block to polyspermy also develops.

Egg-Induced Modifications of the *Zona Pellucida*

Effects of Biologically Active Phorbol Diesters and a Diacylglycerol

Recent studies have centered on the role of protein phosphorylation(s) catalyzed by the calcium- and phospholipid-dependent protein kinase, protein kinase C (PK-C) in regulating exocytotic processes (Nishizuka, 1984; Summers and Creutz, 1985; Takai et al., 1984). Since the cortical granule reaction in mammalian eggs is a poorly understood process and is thought to involve an exocytotic process, we set out to examine the role of mouse egg PK-C in the early events of the fertilization process.

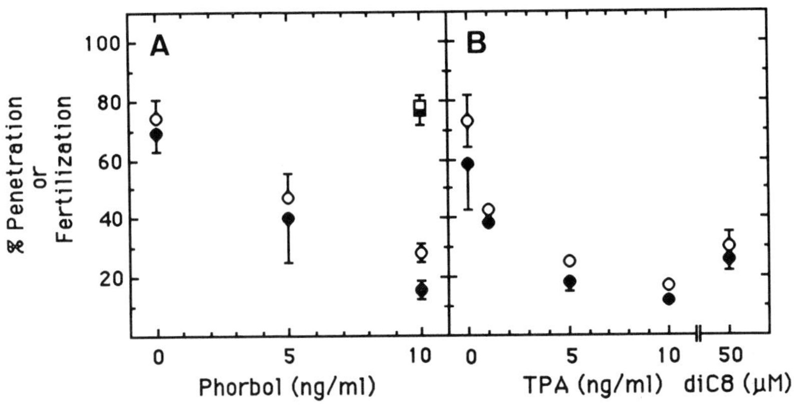

Figure 1. Effects of phorbol diesters and a diacylglycerol on sperm penetration (open symbols) and fertilization (closed symbols). (A) 4β-phorbol-12,13-didecanoate, circles; 4α-phorbol-12,13-didecanoate, squares. (B) TPA. (Reprinted from Endo et al. (1987a) by permission of Academic Press, Inc.)

Treatment of eggs with biological active phorbol diesters, which activate PK-C, resulted in a concentration-dependent inhibition of both sperm penetration and fertilization (Fig. 1) (Endo et al., 1987a). In addition, the more natural activator of PK-C, diacylglycerol (dioctanyl glycerol), inhibited both sperm penetration and fertilization. In contrast, the phorbol diester, 4α-phorbol 12,13-didecanoate, which does not activate PK-C, did not inhibit either sperm penetration or fertilization. Thus, although we did not demonstrate directly that mouse eggs possess PK-C, it was likely that the inhibitory effects on sperm penetration and fertilization elicited by these agents was mediated by a process involving PK-C. It should be noted that the PK-C activators produced these inhibitory effects by acting on the egg and not on the sperm or *zona*. Sperm treated with biologically active phorbol diesters were motile, bound to *zonae*, underwent the acrosome reaction, and penetrated and fertilized eggs to the same extent as untreated sperm. Moreover, isolated *zonae* treated with biologically active phorbol diesters bound and acrosome reacted sperm to the same extent as untreated isolated *zonae* (Endo et al., 1987a).

Table 1
Effect of Treatment of Eggs with PK-C Activators on Sperm Binding[a]

		# Sperm Bound/Zona[b]					
Experiment	Time (h)	Control	4α-PDD	4β-PDD	TPA	diC8	2-PN
I	1	19	11	13			
	2	3	7	5			
	3	2	2	2			
II	1	12	20	22			
	2	2	10	6			
	3	4	11	10			
III	1	13			14	15	
	2	12			11	13	
	3	5			4	4	
IV[c]	1	10			8		1
	2	8			10		2
	3	4			3		1

[a]Reprinted from Endo et al. (1987a) by permission of Academic Press, Inc.
[b]Data are expressed as the average number of sperm bound per *zona*, where five *zonae* were analyzed.
[c]Fertilized eggs with two pronuclei (2-PN) were challenged with capacitated sperm and binding to *zonae* determined.

Inhibition of sperm penetration and fertilization in phorbol diester-treated eggs was due to a step subsequent to sperm binding, since such treated eggs bound similar numbers of sperm as compared to untreated eggs (Table 1). It was likely that the phorbol esters caused an egg-induced modification of the *zona* that prevented sperm penetration. Eggs treated with biologically active phorbol diesters in which the zona was removed (*zona*-free egg) and then challenged with capacitated sperm were fertilized to the same extent as *zona*-free untreated eggs (Endo et al., 1987a). (In a given population of capacitated mouse sperm about 10–20% of the sperm have undergone a spontaneous acrosome reaction; it is these acrosome-reacted sperm that fertilize a *zona*-free egg.)

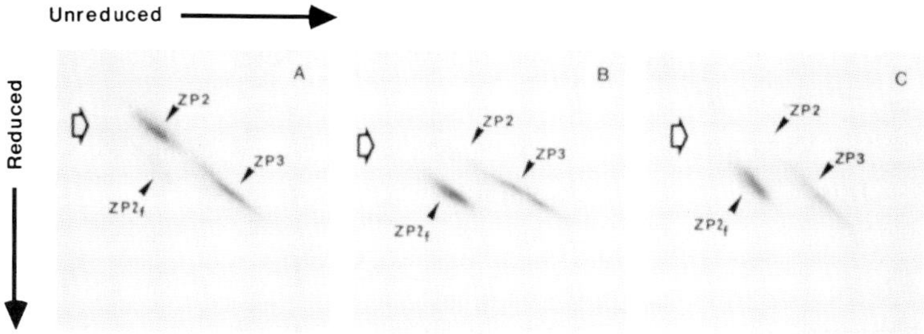

Figure 2. Autoradiograms of *zonae* isolated from (A) untreated eggs, (B) TPA-treated eggs, and (C) 2-cell embryos. The open arrowhead indicates the position of ZP1 that was visible on the original autoradiograms but did not reproduce well. (Reprinted from Endo et al. (1987a) by permission of Academic Press, Inc.)

Two-dimensional reduction gel electrophoresis of iodinated *zonae* obtained from phorbol diester-treated eggs confirmed that the *zona* was modified in response to these PK-C activators (Endo et al.,1987a) (Fig.2). In this procedure, radio-labeled *zonae* are subjected to electrophoresis in the first dimensional under non-reducing conditions (Bleil and Wassarman, 1980b). Thus, ZP1, 2, and 3 migrate with mobilities that correspond to apparent molecular weights of 200,000, 120,000, and 83,000 daltons, respectively. The second dimension is run under reducing conditions that cleave the intermolecular disulfide bonds present in ZP1, so that it migrates with an apparent molecular weight of 100,000 daltons, and ZP2 and ZP3 still migrate with apparent molecular weights of 120,000 and 83,000 daltons, respectively (Fig. 2A). Fertilization results in the conversion of ZP2 to $ZP2_f$, with the corresponding shift in molecular weight to 90,000 daltons. ZP1 and ZP3 show no apparent change in their molecular weights (Fig. 2C). *Zonae* obtained from eggs treated with biologically active phorbol diesters or a diacylglyerol manifest the conversion of ZP2 to $ZP2_f$ similar to that observed in *zonae* obtained from 2-cell embryos (Fig. 2B); eggs treated with a biologically inactive phorbol diester (4α-phorbol 12,13-didecanoate) do not manifest this change.

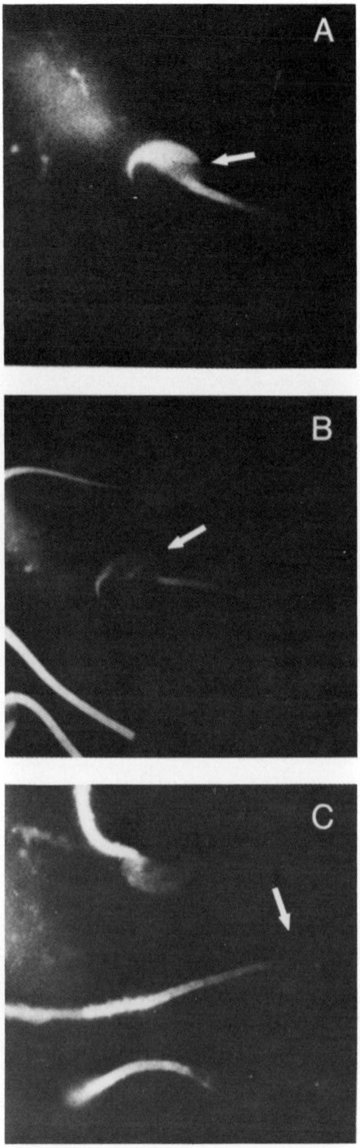

Figure 3. Epifluorescent photomicrographs of sperm at different stages of the acrosome reaction, as observed by the chlortetracycline assay. (A) B-pattern. (B) S-pattern. (C) AR-pattern. (Reprinted from Endo et al. (1987a) by permission of Academic Press, Inc.)

Although these results indicated that activators of PK-C stimulate an egg-induced modification of the *zona*, the potential change in the properties of ZP3 were of initial interest, since ZP3 mediates initial sperm binding. As described previously, although such treated eggs bind sperm to a similar extent as untreated eggs, the bound sperm fail to penetrate and fertilize the egg. A possible explanation for this observation is that sperm bound to *zonae* from phorbol diester-treated eggs fail to undergo the acrosome reaction.

Some of the changes that culminate in the acrosome reaction can be monitored by changes in staining patterns using the fluorescent probe chlortetracycline (Fig. 3) (Saling and Storey, 1979). Three major fluorescent staining patterns have been characterized with this assay. The B-pattern of capacitated sperm is correlated with acrosome intact sperm, as assessed by transmission electron microscopy (Florman and Storey, 1982). The S-pattern represents an intermediate stage and appears prior to completion of the acrosome reaction. This pattern correlates with loss of the ability of sperm to maintain a transmembrane pH gradient (Lee and Storey, 1985), and may occur as a result of initial stages of vesiculation due to fusion of the outer acrosomal and plasma membranes. The AR-pattern corresponds to acrosome-reacted sperm, as determined by transmission electron microscopy (Saling et al., 1979).

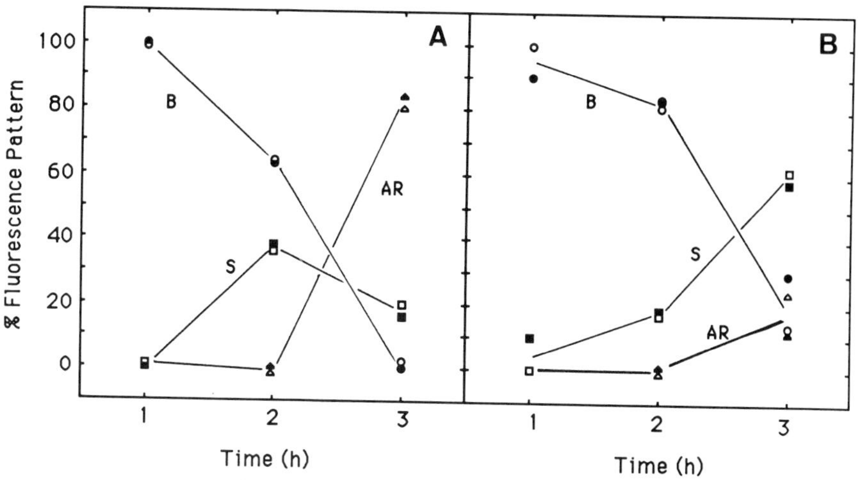

Figure 4. Effects of phorbol diester treatment of *zona*-intact eggs on the acrosome reaction of sperm bound to the *zona* (A). Untreated (open) or 4α-phorbol-12,13-didecanoate-treated (closed symbols) eggs. (B) 4β-phorbol-12,13-didecanoate-treated (open symbols) or dioctanoyl acylglycerol-treated (closed symbols) eggs. (Reprinted from Endo et al. (1987a) by permission of Academic Press, Inc.)

Sperm treated with phorbol diesters display the same series of changes in chlortetracycline fluorescence as do untreated sperm (Endo et al., 1987a). Thus, it was possible to compare the kinetics of the acrosome reaction of bound sperm to *zonae* obtained from either untreated or phorbol diester-treated eggs (Fig. 4). Sperm bound to *zonae* of untreated eggs or eggs treated with a biologically inactive phorbol diester underwent B to S to AR transitions with similar kinetics. In contrast, sperm bound to the *zonae* of eggs treated with either a biologically active phorbol diester or diacylglycerol underwent the B to S transition but did not progress into the AR pattern, since they accumulated in the S pattern. Thus, although the bound sperm initiated the acrosome reaction, they failed to complete it.

Fertilization is normally accompanied by a loss of both activities associated with ZP3. Results of these experiments indicated an apparent dissociation in intact *zonae* of sperm receptor activity from acrosome-inducing activity. It was possible, however, that rather than a modification of ZP3 being solely responsible for the dissociation of these two biological activities of ZP3 the structural integrity of the *zona* was necessary to observe these changes. Therefore, ZP3 was isolated from untreated eggs and phorbol diester-treated eggs and tested for its two biological activities.

Sperm competition binding assays revealed that, as expected, ZP3 isolated from phorbol diester-treated eggs possessed similar amounts of sperm receptor activity as compared to ZP3 isolated from untreated eggs (Fig. 5) (Endo et al., 1987b). The competition by ZP3 for sperm binding to

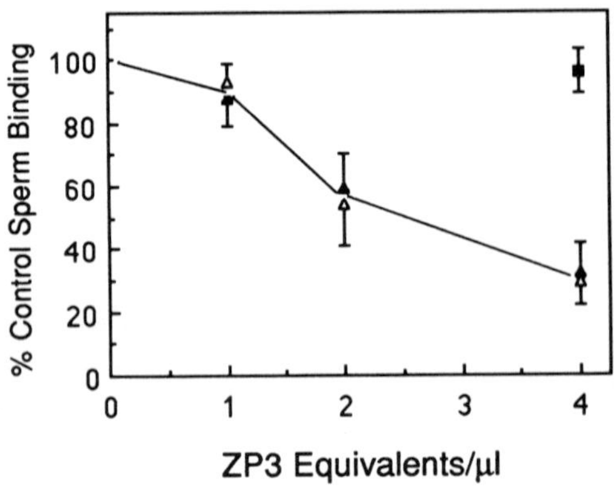

Figure 5. Sperm receptor activity of ZP3 isolated from untreated (▲), TPA-treated (△) eggs, and 2-cell embryos (■). Sperm competition assays were performed in which increasing amounts of ZP3 isolated from the aforementioned eggs was added and its effect on the binding of sperm to *zona*-intact eggs was measured. (Reprinted from Endo et al. (1987a) by permission of Academic Press, Inc.)

zona-intact eggs is specific, since ZP3 isolated from 2-cell embryos, which does not possess sperm receptor activity (Bleil and Wassarman, 1980c), did not possess sperm receptor activity (Fig. 5). When examined for the ability to induce the acrosome reaction, ZP3 isolated from untreated eggs possessed the ability to induce the B to S to AR transitions, as monitored by the chlortetracycline assay (Fig. 6). In contrast, as was the case with structurally intact *zonae* isolated from phorbol diester-treated eggs, ZP3 isolated from these eggs did not induce a full acrosome reaction, but rather, sperm accumulated in the S pattern (Fig. 6).

Although previous studies indicated that ZP3 isolated from 2-cell embryos did not induce the acrosome reaction, the methods used in these studies assayed an end point, i.e., the completion of the acrosome reaction, and would not detect intermediates in this process (Bleil and Wassarman, 1983). When tested with the chlortetracycline assay, sperm incubated with ZP3 isolated from 2-cell embryos did not even undergo the B to S transition (Fig. 6).

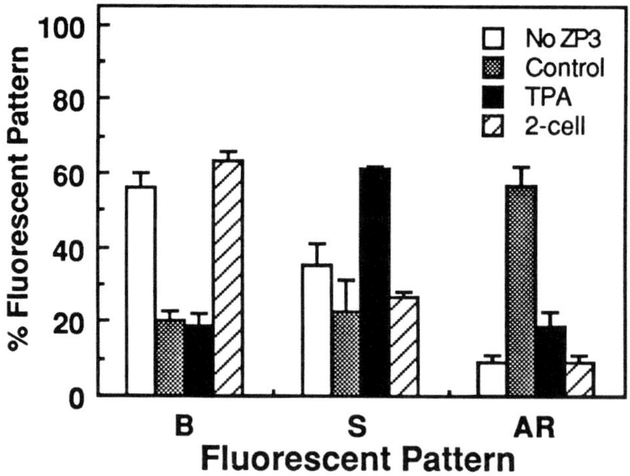

Figure 6. Distribution of chlortetracycline fluorescent patterns of sperm incubated with no ZP3, or ZP3 isolated from untreated eggs, TPA-treated eggs, or 2-cell embryos. It should be noted that a fraction of sperm undergo a spontaneous B to S transition (Lee and Storey, 1985). In addition, in a given population of capacitated sperm, there exists a subpopulation of about 10–20% that have undergone the acrosome reaction. The differences in the distribution of the chlortetracycline patterns between sperm incubated in the absence of ZP3 and ZP3 isolated from 2-cell embryos are not significant, whereas the differences in the distribution of both the S and AR patterns of sperm isolated in the presence of ZP3 isolated from either untreated or TPA-treated eggs are significant. (Reprinted from Endo et al. (1987a) by permission of Academic Press, Inc.)

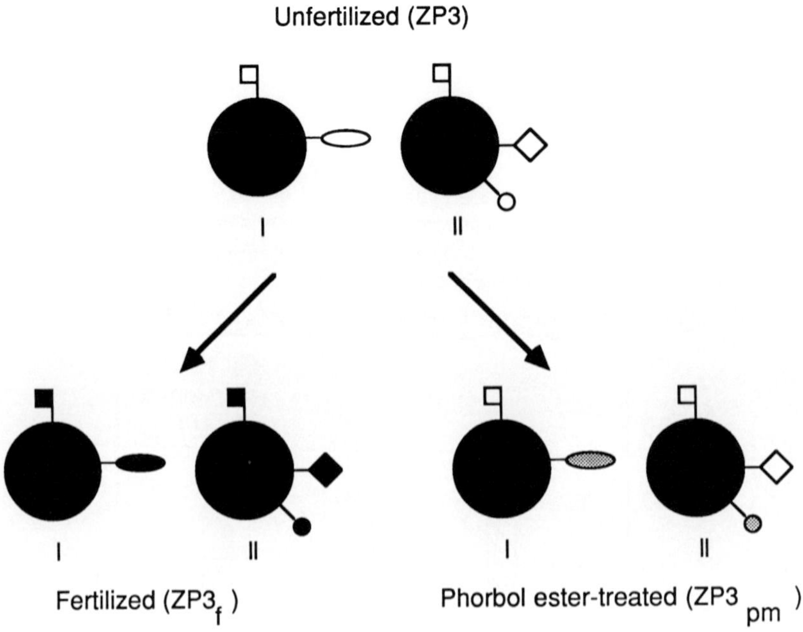

Figure 7. Schematic representation of changes in ZP3 following fertilization or phorbol diester treatment. The figure depicts two possibilities that currently cannot be distinguished. Case I represents a model in which one component (ellipse) of ZP3 is responsible for both the B to S and S to AR transitions of the acrosome reaction. Case II represents a model in which one component of ZP3 is responsible for the B to S transition (diamond) and another component is responsible for the S to AR transition (circle). In both cases, the square denotes the sperm receptor activity component of ZP3. $ZP3_f$ represent ZP3 isolated from fertilized eggs and $ZP3_{pm}$ represents ZP3 isolated from phorbol diester-treated eggs that displays a partial modification (pm) of the acrosome reaction-inducing activity. (Reprinted from Endo et al. (1987a) by permission of Academic Press, Inc.)

Thus, treatment of eggs with activators of protein kinase C results in an egg-induced modification of the *zona* such that there is a dissociation of the sperm receptor activity from the acrosome reaction-inducing activity of ZP3. Presumably, differences in ZP3 obtained from untreated and phorbol diester-treated eggs reflect those portions of the molecules that participate in the acrosome reaction (Fig. 7). Studies examining biochemical differences in ZP3 obtained from untreated and phorbol diester-treated eggs, therefore, should facilitate analysis of those portions of ZP3 that are involved in inducing the acrosome reaction. The use of *zonae* from phorbol diester-treated eggs should also facilitate studies on

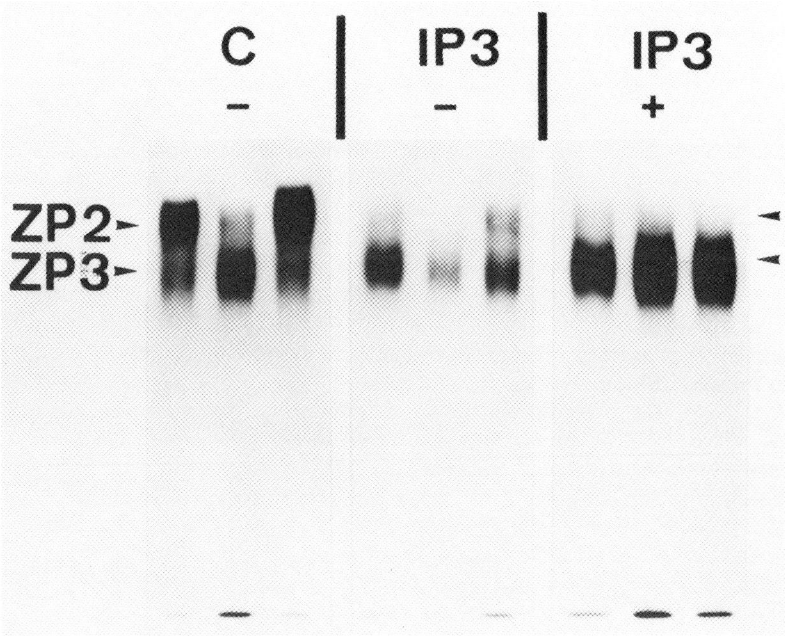

Figure 8. Autoradiograms of individual *zonae* isolated from either control injected or IP_3-injected eggs. (A). Vehicle-injected eggs. (B). IP_3-injected eggs that did not activate. (C). IP_3-injected eggs that did activate. In B and C, note the consistent loss of ZP2.

the mechanism(s) of the acrosome reaction, since it is now possible to use these *zonae* to study independently the B to S transition from the S to AR transition. Such an experimental system may be of great value in determining the biochemical correlates of the S pattern, which has characteristics of an intermediate stage prior to the completion of the acrosome reaction.

Effects or Microinjected Inositol-1,4,5-trisphosphate

Phosphatidyl inositol turnover, which has been implicated as a response of the egg to fertilization in lower species (Turner et al., 1984, 1986; Whitaker and Irvine, 1984), generates two second messengers, i.e., diacylglycerol, which activates PK-C, and inositol-1,4,5-trisphosphate (IP_3), which is involved in releasing calcium from intracellular stores (Berridge, 1984; Berridge and Irvine, 1984). Our previous experiments using PK-C activators would in principle only activate one half of this pathway. Accordingly, we examined the effect of microinjecting IP_3 into mouse eggs on egg activation, modifications of the *zona*, and sperm receptor and acrosome inducing activities of the *zona*.

About 15% of eggs injected with IP_3 at a final concentration of about 4 μM underwent activation as assessed by emission of the second polar body within 1.5 h after injection; no eggs injected with the vehicle solution underwent activation. When *zonae* of these microinjected eggs were isolated, radioiodinated, and subjected to one dimensional gel electrophoresis under reducing conditions, about 85% of those that did not activate and about 95% of those that did activate manifested the loss of ZP2 and its presumable conversion to $ZP2_f$ (Fig. 8). In contrast, only about 30% of eggs injected with the vehicle manifested this conversion, and in most instances the conversion to $ZP2_f$ was not complete (Fig. 8).

Although analysis of the number of sperm bound to such IP_3-injected eggs revealed that at the initial time point fewer sperm were bound relative to vehicle-injected eggs (Table 2), there was no statistical difference in the kinetics of the acrosome reaction of bound sperm, as determined by the chlortetracycline assay, between IP_3- and vehicle-injected eggs (Table 2). We believe that the initial lower level of sperm binding does not reflect any change in sperm receptor activity, i.e., ZP3 in *zonae* of these IP_3-injected eggs possesses both sperm receptor and acrosome-inducing reaction activities, for the following reason. If ZP2 mediates binding of acrosome reacted sperm and $ZP2_f$ cannot interact with acrosome-reacted sperm, then although sperm can bind to IP_3-injected eggs and undergo the acrosome reaction, they cannot establish the secondary binding interaction with ZP2, which has been converted to $ZP2_f$. Hence, the sperm that were initially bound subsequently dissociated from the (Fig. 9). Thus, at the early time point, which is one hour after sperm binding, we would expect a lower level of binding.

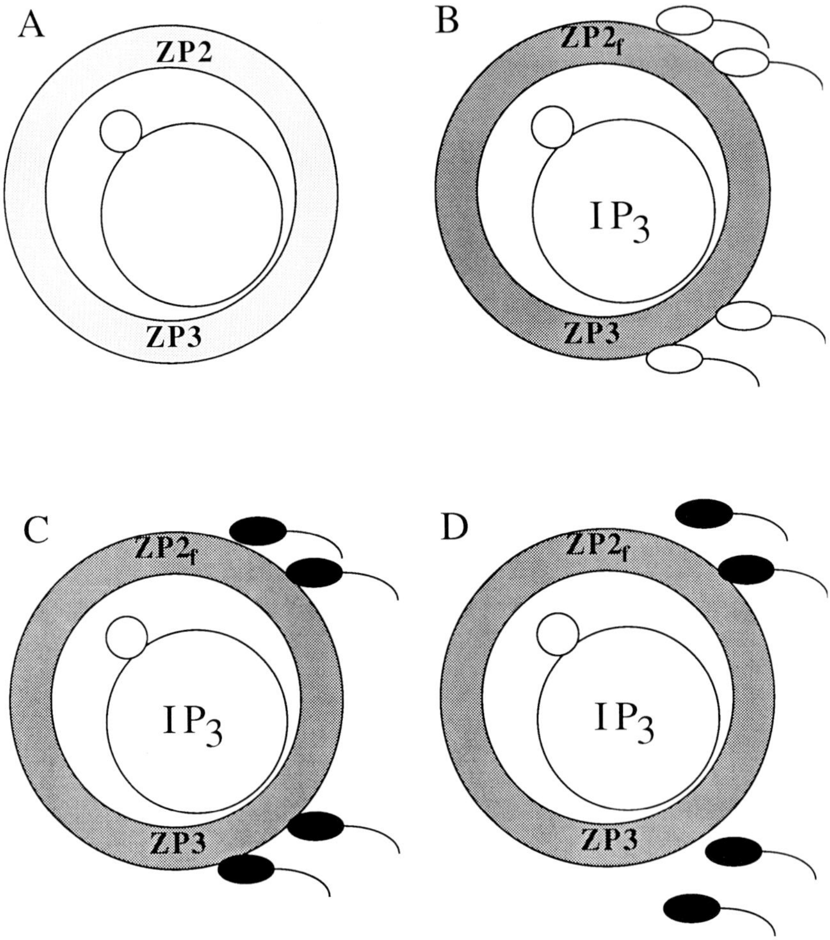

Figure 9. Schematic representation of effect of microinjected IP_3 on sperm binding. A. Unfertilized egg with ZP3 that possesses both sperm receptor activity and the ability to induce the acrosome reaction and ZP2 that binds acrosome-reacted sperm. B. Injection of IP3 results in a modification of the *zona*, as depicted by its darker shade, in which ZP3 remains intact but ZP2 is converted to $ZP2_f$; which does not interact with acrosome-reacted sperm. Acrosome-intact sperm (unshaded) bind to the *zona* via ZP3. C. ZP3 induces the acrosome reaction of bound sperm (dark sperm). D. The acrosome-reacted sperm, which do not interact with ZP3, cannot interact with $ZP2_f$ and therefore detach from the *zona*.

Table 2
Effect of Microinjection of Inositol 1,4,5-Trisphosphate into Mouse Eggs on Sperm Binding and the Acrosome Reaction[a]

Experiment	Time (h)	# Sperm Bound[c]	% Fluorescent Pattern[b]		
			B	S	AR
Control	1	12.1±3.1	91	9	0
	2	3.9±2.5	61	32	7
	3	5.7±1.1	2	35	63
IP3	1	5.5±4.0	84	12	0
	2	4.4±2.1	42	44	14
	3	4.1±3.6	13	45	43

[a]Data represent either the mean ±S.D. or mean of five individual experiments.
[b]Patterns analyzed by the chlortetracycline assay.
[c]Data are expressed as the average number of sperm bound per *zona*, where five *zonae* were analyzed.

Results of these studies demonstrate the first dissociation of functional changes between ZP2 and ZP3 in structurally intact *zonae* and should facilitate studies of the biological role of changes in ZP2. To date, the role of ZP2 and the consequence of its conversion to $ZP2_f$ have been based on studies of the purified molecule.

A possible explanation for the differential effects of PK-C activators and IP_3 on these egg-induced modifications of the mouse egg *zona pellucida* may be a consequence of heterogeneity in the cortical granule population. Electron microscopy studies reveal the existence of a light and dark staining populations of cortical granules (Nicosia et al., 1977). Perhaps the treatments described here result in the selective release of the contents of a subpopulation of cortical granules that are capable of only performing certain aspects of the *zona* modifications, which *in toto* constitute the *zona* reaction. Consistent with this hypothesis is the observation that eggs treated with biologically active phorbol diesters, but not inactive ones, display a partial reduction in the number of lens culinaris agglutinin-staining granules, which are presumptive cortical granules, when compared to ionophore-activated or fertilized eggs (Ducibella, Kopf, and Schultz, unpublished results). Additional studies are required to verify this hypothesis.

ACKNOWLEDGEMENTS

Research described in this manuscript was supported by grants from the NIH (HD 19096 to G.S.K. and HD 18604 to R.M.S. and HD 22732 to G.S.K. and R.M.S.) and grants from the Mellon Foundation (G.S.K.) and University Research Fund (R.M.S.). Y.E. was supported by the Rockefeller

Foundation. G.S.K. and R.M.S. would like to thank Jeff Bleil for stimulating conversions regarding the role of ZP2 and $ZP2_f$ in secondary binding of sperm and *zona* block.

REFERENCES

Barros, C., and Yanagimachi, R. (1971). Induction of the zona reaction in golden hamster eggs by cortical granule material. Nature (London) 233, 268–269.

Barros, C., and Yanagimachi, R. (1972). Polyspermy-preventing mechanisms in the golden hamster egg. J. Exp. Zool. 180, 251–266.

Berridge, M.J. (1984). Inositol triphosphate and diacylglycerol as second messengers. Biochem. J. 220, 345–360.

Berridge, M.J., and Irvine, R.F. (1984). Inositol triphosphate, a novel second messenger in cellular signal transduction. Nature (London) 312, 315–321.

Bleil, J.D., and Wassarman, P.M. (1980a). Structure and function of the *zona pellucida*: Identification and characterization of the proteins of three mouse oocyte's *zona pellucida*. Dev. Biol. 76, 185–202.

Bleil, J.D., and Wassarman, P.M. (1980b). Synthesis of *zona pellucida* proteins by denuded and follicle-enclosed mouse oocytes during culture *in vitro*. Proc. Natl. Acad. Sci. USA 77, 1029–1033.

Bleil, J.D., and Wassarman, P.M. (1980c). Mammalian sperm-egg interaction: Identification of a glycoprotein in mouse egg *zonae pellucidae* possessing receptor activity for sperm. Cell 20, 873–882.

Bleil, J.D., Beall, C.F., and Wassarman, P.M. (1981). Mammalian sperm-egg interaction: Fertilization of mouse eggs triggers modification of the major *zona pellucida* glycoprotein, ZP2. Dev. Biol. 86, 189–197.

Bleil, J.D., and Wassarman, P.M. (1983). Sperm egg interactions in the mouse: Sequence of events and induction of the acrosome reaction by a *zona pellucida* glycoprotein. Dev. Biol. 95, 317–324.

Bleil, J.D., and Wassarman, P.M. (1986). Autoradiographic visualization of the mouse egg's sperm receptor bound to sperm. J. Cell Biol. 102, 1363–1371.

Endo, Y., Schultz, R.M., and Kopf, G.S. (1987a). Effects of phorbol esters and a diacylglycerol on mouse eggs: Inhibition of fertilization and modification of the *zona pellucida*. Dev. Biol. 119, 199–209.

Endo, Y., Mattei, P., Kopf, G.S., and Schultz, R.M. (1987b). Effects of a phorbol ester on mouse eggs: Dissociation of sperm receptor activity from acrosome reaction-inducing activity of the mouse *zona pellucida* protein, ZP3. Dev. Biol. 123, 574–577.

Florman, H.M., and Storey, B.T. (1982). Mouse gamete interactions: The *zona pellucida* is the site of the acrosome reaction leading to fertilization *in vitro*. Dev. Biol. 91, 121–130.

Florman, H.M., Bechtol, K.B., and Wassarman, P.M. (1984). Enzymatic dissection of the functions of the mouse egg's receptor for sperm. Dev. Biol. 106, 243–255.

Florman, H.M., and Wassarman, P.M. (1985). O-linked oligosaccharides of mouse egg ZP3 account for its sperm receptor activity. Cell 41, 313–324.

Greve, J.M., and Wassarman, P.M. (1985). Mouse egg extracellular coat is a matrix of interconnected filaments possessing a structural repeat. J. Mol. Biol. 181, 253–264.

Lee, M.A., and Storey, B.T. (1985). Evidence for plasma membrane impermeability to small ions in acrosome-intact mouse spermatozoa bound to mouse *zonae pellucidae*, using an aminoacridine fluorescent probe: Time course of the *zona*-induced acrosome reaction monitored by both chlortetracycline and pH probe fluorescence. Biol. Reprod. 33, 235–246.

Nicosia, S.V., Wolf, D.P., and Masato, I. (1977). Cortical granule distribution and cell surface characteristics in mouse eggs. Dev. Biol. 57, 56–74.

Nishizuka, Y. (1984). The role of protein kinase C in cell surface signal transduction and tumour promotion. Nature (London) 308, 693–698.

Philpott, C.C., Ringuette, M.J., and Dean, J.D. (1987). Oocyte-specific expression and developmental regulation of ZP3, the sperm receptor of the mouse *zona pellucida*. Dev. Biol. 121, 568–575.

Saling, P.M., and Storey, B.T. (1979). Mouse gamete interactions during fertilization *in vitro*: Chlortetracycline as fluorescent probe for mouse sperm acrosome reaction. J. Cell Biol. 83, 544–555.

Saling, P.M., Sowinski, J., and Storey, B.T. (1979). An ultrastructural study of epididymal mouse spermatozoa binding to *zonae pellucidae in vitro*: Sequential relationship to the acrosome reaction. J. Exp. Zool. 209, 229–238.

Shimizu, S., Tsuji, M., and Dean, J. (1983). *In vitro* biosynthesis of three sulfated glycoproteins of murine zonae pellucidae by oocytes grown in follicle culture. J. Biol. Chem. 258, 5858–5863.

Summers, T.A., and Creutz, C.E. (1985). Phosphorylation of a chromaffin granule-containing protein by protein kinase C. J. Biol. Chem. 260, 2437–2443.

Szollsi, D. (1967). Development of cortical granules and the cortical reaction in rat and hamster eggs. Anat. Rec. 159, 431–446.

Takai, Y., Kikkawa, U., Kaibuchi, K., and Nishizuka, Y. (1984). Membrane phospholipid metabolism and signal transduction for protein phosphorylation. Adv. Cyclic Nuc. Prot. Phos. Res. 18, 119–154.

Turner, P.R., Sheetz, M.P., and Jaffe, L.A. (1984). Fertilization increases the polyphosphoinositide content of sea urchin eggs. Nature (London) 310, 414–415.

Turner, P.R., Jaffe, L.A., and Fein, A. (1986). Regulation of cortical granule exocytosis in sea urchin eggs by inositol 1,4,5-trisphosphate and GTP-binding protein. J. Cell Biol. 102, 70–76.

Whitaker, M., and Irvine, R.F. (1984). Inositol 1,4,5-trisphosphate microinjection activates sea urchin eggs. Nature (London) 312, 636–639.

Wolf, D.P., and Hamada, M. (1977). Induction of zonal and oolemmal blocks to sperm penetration in mouse eggs with cortical granule exudate. Biol. Reprod. 17, 350–354.

Cellular Factors in Development and Differentiation:
Embryos, Teratocarcinomas, and Differentiated Tissues, pages 93–104

Organization of homeodomain proteins

Michael Kessel, Mathias Fibi and Peter Gruss

Department of Molecular Cell Biology,
Max Planck Institute of Biophysical Chemistry,
Am Fassberg, 3400 Göttingen, FRG

INTRODUCTION

In Drosophila melanogaster a number of genes have been identified which play key roles in the generation of pattern formation during embryonic development (for review see Scott et al., 1986). Genetic approaches have been used to isolate numerous of these Drosophila genes and establish their organization and structure. It turned out that a subset of these genes possesses a conserved DNA segment of 183 bp in length, termed the "homeo box", which encodes a part of the gene product termed the "homeodomain" (Gehring 1985). The high degree of conservation is not restricted to the Drosophila homeo box genes but could also be demonstrated to extend to earthworms, leeches, beetles, shrimps, sea urchins, tunicates, amphioxus, frogs, chicken, mice and humans, while bacteria, fungi, sea anemonies, planaria, tapeworms, nematodes and squid appear to lack such sequences (McGinnis 1985). Using the Drosophila sequences as molecular probes numerous genes have now been isolated from many other species. Most of these contain a homeo box highly similar to the prototype Drosophila homeo box found in the homeotic gene Antennapedia, which is referred to as a class I box. However, it is clear that at least in Drosophila there are other classes of boxes, which are only 50% or less related to the Antp box. Such classes are found in the caudal gene (Mlodzik et al., 1987), the engrailed gene (Poole et al., 1985), the even-skipped gene (McDonald et al., 1986), the paired gene (Frigerio et al., 1986) and the bicoid gene (Bopp et al., 1986), all of which seem to represent independent classes of homeo boxes. It remains to be seen if these types of boxes can also be found in the genomes of other species.

The function of many Drosophila homeo box genes was to a certain degree clarified by investigating the corresponding mutants. Thus, in many cases it could be quite precisely determined at which point they are involved in the generation of biological structures and how they interact in a network with other genes (Scott et al., 1986). Since a genetic approach is not feasible in higher organisms in no case could such a deep functional understanding be achieved for their homeo box genes. The characterization of these genes therefore has emphasized structural aspects and concentrated on descriptions of gene expression in different tissues.

The analysis of transcription of homeo box genes in humans, mice and Xenopus demonstrated that their expression is spatially and temporally restricted. Most notably they show high levels of expression in certain embryonal tissues, a finding which was expected if these genes played a role in the regulation of vertebrate development.

In a second phase of study the attention had to focus on the structure of the gene products, the homeodomain containing proteins, which could be analyzed via cDNA clones. By now the complete primary structures of more than ten homeodomain proteins can be predicted from cDNA sequences. These include the Drosophila proteins Antennapedia (Antp; Schneuwly et al., 1986), Deformed (Dfd; Regulski et al., 1987), fushi tarazu (ftz; Laughon et al., 1984), engrailed (en; Poole et al., 1985), invected (inv; Colemon et al., 1987), caudal (cad; Mlodzik et al., 1987) and even-skipped (eve; McDonald et al., 1986); the human proteins C13 (Mavilio et al., 1986), C1 and C8 (Simeone et al., 1987); the Xenopus protein Xhox 1 (Harvey et al., 1985); and the mouse proteins Hox 2.1 (Krumlauf et al., 1987), Hox 1.1 (Kessel et al., 1987) and Hox 1.3 (Odenwald et al., 1987; Fibi et al., 1987). Table I lists published proteins and some relevant data from the original literature. In the following paragraphs we compare especially those proteins with a class I domain in order to find general principles in the organization of these molecules and to point out highly conserved regions. An understanding of the structural organization may help in

	Organism	number of amino acids	Mr	Hexapeptide	related protein	reference
Ant	Drosophila	378	43000	+	-	Schneuwly et al., 1986
ftz	Drosophila	413	45000	-	-	Laughon et al., 1984
Dfd	Drosophila	586	63000	+	Xhox 1A, C13, (Hox 1.4)	Regulski et al., 1987
Xhox-1	Xenopus	232	26000	+	Dfd, C13, (Hox 1.4)	Harvey et al., 1985
Hox 1.1	Mouse	229	25000	+	C1, Hox 2.3, MM3	Kessel et al., 1987
Hox 1.3	Mouse	269	30000	+	Hox 2.1	Fibi et al., 1987
Hox 2.1	Mouse	269	29000	+	Hox 1.3	Krumlauf et al., 1987
C 13	Human	255	28000	+	Xhox 1A, Dfd, (Hox 1.4)	Mavilio et al., 1986
C 1	Human	217	24000	+	Hox 1.1, Hox 2.3, MM3	Simeone et al., 1987
C 8	Human	153	17000	+	AC1	Simeone et al., 1987

Table 1: Homeodomain proteins
Relationships between proteins are tentatively assigned on the basis of sometimes only limited sequence homology, see text and figures.

finding a function for these proteins. In comparing these proteins two aspects have to be kept in mind:

Firstly, comparing homeodomain proteins of different species may lead to the identification of proteins, which have conserved many of their structural features in order to fulfill a specific comparable function. A possible example for such an interspecies group are the proteins Dfd, Xhox-1, Hox 1.4 and C13, which may represent the corresponding proteins of Drosophila, Xenopus, mice and humans, respectively. Secondly, the comparison of different homeodomain proteins of a single species may point out proteins which evolved by gene duplication. A quite probable example for such a case is the murine pair Hox 1.3 and Hox 2.1 (Fibi et al., 1987).

For the purpose of this discussion we have divided a prototype class I homeodomain protein into six regions (A-F), which will at first be discussed separately (Fig. 1).

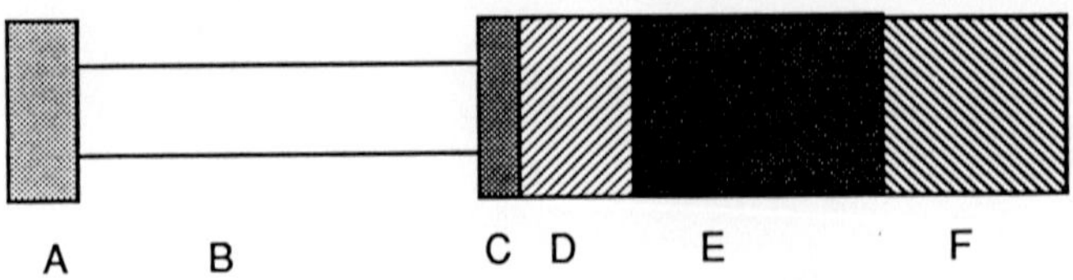

Figure 1. Prototype homeodomain protein

A) A general feature of the N-termini of many class I homeodomain proteins is the sequence Met-Ser-Ser. Furthermore, the first 20 amino acids seem to be subject to evolutionary pressure as exemplified by a good degree of conservation between several proteins. Comparing the class I proteins it is obvious that some sequence elements are often found like Met-Ser-Ser followed by one or two aromatic amino acids, or the peptide Ser-Lys-Tyr. Furthermore, there are more closely related proteins like Hox 2.1 and Hox 1.3, Hox 1.1 and C1 or Xhox 1A and C13. We interpret these observations as the indication for a loosely conserved domain, which may be highly conserved if the proteins belong to the same subfamily.

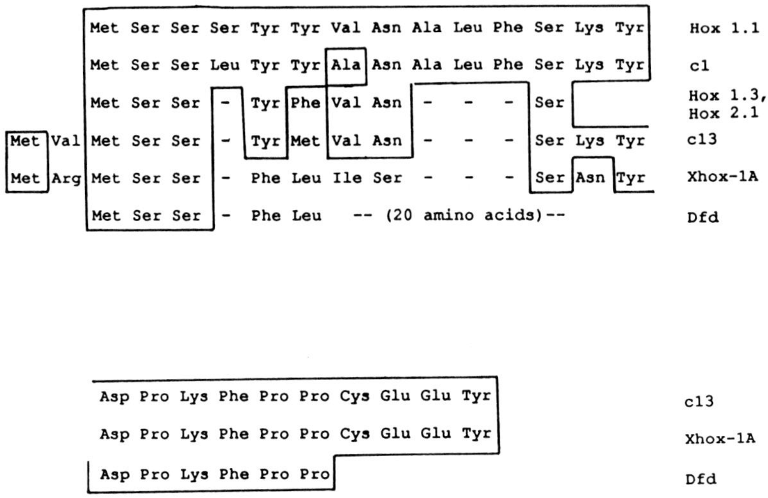

Figure 2. Comparison of N-terminal sequences

B) In the region after the N-terminal 20 amino acids, extending roughly over the first half of the protein, no conservation is observed in any of the homeodomain proteins, with the exception of the pair Hox 1.3 and Hox 2.1, where the N-terminal conserved region extends further, before divergence starts. Of course it remains unclear if this region is of lesser importance and thus not conserved, or if this region confers the individual and specific function to the protein.

C) Very close to the homeodomain there is a second conserved domain consisting of a hexapeptide which is common to all class I homeodomain proteins described so far with the exception only of the Drosophila segmentation gene fushi tarazu (Laughon et al., 1984). The consensus for this sequence reads Ile (or another alphatic amino acid) - Tyr-Pro-Trp-Met-Arg/Lys (Fig. 3, 4, 5). Since in all cases there is a splice site right before the homeo box this hexapeptide is always encoded on an exon preceding the homeo box.

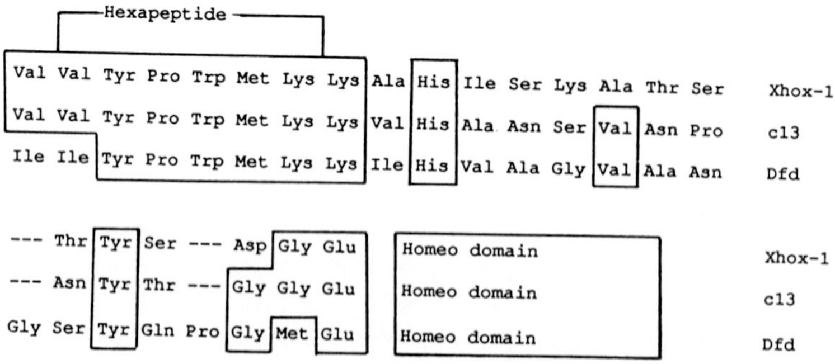

Figure 3. Interspecies conservation of regions C and D.

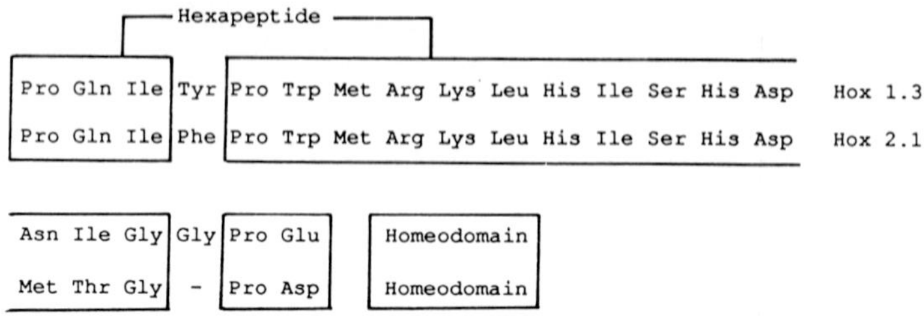

Figure 4. Conservation of regions C and D between two related murine proteins.

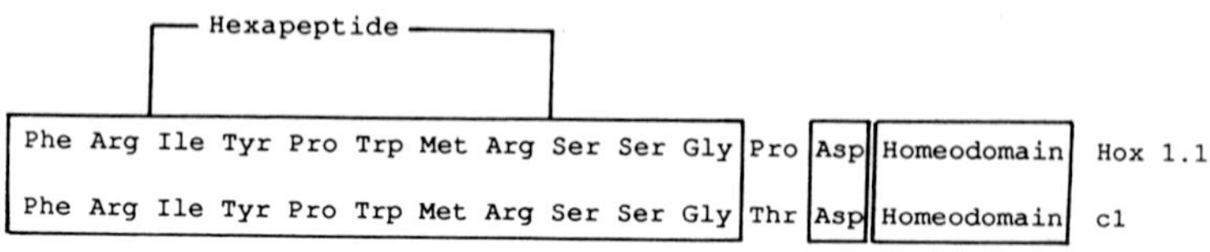

Figure 5. Conservation of regions C and D between related proteins from mice and humans.

D) The region between the hexapeptide (C) and the start of the homeodomain (E) varies in length between 5 and 17 nonrelated amino acids. Analysis of the two groups mentioned above shows a limited similarity in length and some conserved residues for the Dfd group (Fig. 3) and the Hox 1.3/2.1 pair (Fig. 4), and the Hox 1.1/C1 pair (Fig. 5).

E) The strong conservation of the homeodomain is the basis for grouping these proteins together and has been documented and discussed in detail in the original literature (for review see Gehring 1985). It should briefly be mentioned here that the homeodomain bears loose homology to prokaryotic DNA-binding proteins such as the lac repressor and yeast mating type proteins. In analogy it can be predicted that the homeodomain folds into a helix-turn-helix secondary structure, which is likely to be DNA binding. Other regions for example dimerization points may be present on this domain (P. O'Farrell, pers. commun.).

F) The region directly adjacent to the homeodomain is not generally conserved in primary structure and length, but only between apparently corresponding proteins. Thus, the Dfd group has conserved the 5 amino acids following the homeodomain and then the sequences diverge completely (Fig. 6). On the other hand, the two proteins Hox 1.2 and 1.3 both contain 15 amino acids in this region of which only 4 differ (Fig. 7). Since also partial genomic sequences can be compared regarding region F, a number of other matching genes can be described. Thus ACI (Xenopus; Carrasco et al., 1984) and C8 (human; Simeone et al., 1987) have identical F regions and murine Hox 2.3 (Lonai et al., 1987), MM3 (Xenopus; Müller et al., 1984) and C1 (human; Simeone et al., 1987) are almost identical. There are preliminary indications that this tentative grouping into subfamilies is corroborated by homologies in the regions A and C/D.

Homeodomain	Leu Pro Asn Thr Lys	Xhox-1
Homeodomain	Leu Pro Asn Thr Lys	c13
Homeodomain	Leu Pro Asn Thr Lys	Dfd
Homeodomain	Leu Pro Asn Thr Lys	Hox 1.4

Figure 6. Interspecies conservation of the five amino acids adjacent to the homeodomain.

Leu Lys Ser Met Ser	Met	Ala	Ala	Ala Gly	Gly	Ala Phe	Arg	Pro	Hox 1.3
Leu Lys Ser Met Ser	Leu	Ala	Thr	Ala Gly	Ser	Ala Phe	Gln	Pro	Hox 2.1

Figure 7. Conservation of the complete Carboxy terminus between two highly related murine proteins.

Not appearing in the model (Fig. 1) are the stretches of identical amino acids, which are present in many homeodomain proteins. This is especially evident in several Drosophila proteins, e.g., Dfd, where many such stretches are present, but often they also occur in the proteins of vertebrates. C13 possesses a serine stretch in region F, while in the closely related protein Xhox-1 it is located in region B. Possibly most of the homogenous stretches have been lost during evolution of the Dfd protein leaving only one serine stretch in the vertebrate proteins. This may be one reason for the larger size of the Drosophila proteins. Further noteworthy are the poly glutamic acid stretches that occur in many vertebrate proteins very near or at the carboxy terminus. Thus, the Xenopus protein MM3 (Müller et al., 1984), the mouse protein Hox 2.3 (Lonai et al., 1987) and the human protein C1 (Simeone et al., 1987) contain a region F with 6 consecutive glutamic acid residues. The murine Hox 1.1 protein contains 15 glutamic acid residues at the carboxy terminus (Kessel et al., 1987) and the Xenopus protein ACI and the human protein C8 are also glu-rich in region F (Carrasco et al., 1984; Simeone et al., 1987) An acidic structure at the carboxy terminus is also a feature of other nuclear proteins such as the HMG proteins or the N1/N2 proteins (Kleinschmidt et al., 1986). It may be necessary for the action of these proteins in the nucleus where they may have to interact with basic charges, e.g., from histones or from the homeodomain itself.

Summarizing the available data on class I homeodomain proteins several regions or domains are becoming evident which are subject to evolutionary pressure. These regions are conserved within homeo proteins of one species, as well as between those of different species. It appears that more than the mere 60 amino acids of the homeodomain (region E) are conserved in cases of a closer relationship. Regions upstream and downstream will then form a larger domain C-D-E-F covering a major part of the protein. Region C, located at the N-terminal end of this extended domain may have the function of separating the extended homeodomain from the rest of the molecule by allowing the molecule to bend at this position. The general conservation of the hexapeptide would be consistent with such a general, basic function.

In addition to the large domain C-D-E-F a second domain is represented by the N-terminus. Here we observe general conservation, as well as specific conservation between possibly more closely related pairs or groups of proteins. The two domains A and C-D-E-F are separated by a large region (B) on which no or only little selective pressure is exerted.

Concerning the function of the discussed regions only the DNA-binding capability of the second half of region E, the homeodomain, is known (Desplan et al., 1985). It is possible that the first half of E is responsible for dimerization of the protein and that the hexapeptide (C) is needed for bending the molecule. A deeper understanding of the organization of these molecules will however require much more specific information. Analysis of the different structural features by mutation in vitro and gene transfer may help to elucidate how this family of highly conserved proteins can fulfill specific functions during development and differentiation in higher organisms.

REFERENCES

Bopp D, Burri M, Baumgartner S, Frigerio G, Noll M (1986). Conservation of a large protein domain in the segmentation gene paired and in functionally related genes of Drosophila. Cell 47:1033-1040.

Carrasco AE, McGinnis W, Gehring WJ, De Robertis EM (1984). Cloning of an X.leavis gene expressed during early embryogenesis coding for a peptide region homologous to Drosophila homeotic genes. Cell 37:409-414.

Coleman KG, Poole SJ, Weir MP, Soeller WC, Kornberg T (1987). The invected gene of Drosophila: sequence analysis and expression studies reveal a close kinship to the engrailed gene. Genes & Development 1:19-28.

Desplan C, Theis J, O'Farell PH (1985). The Drosophila developmental gene, engrailed, encodes a sequence-specific DNA binding acitivity. Nature 318:630-635.

Fibi M, Zink B, Kessel M, Colberg-Poley AM, Labeit S, Lehrach H, Gruss P (1987) Coding sequence and expression of the homeo box gene Hox 1.3. Development (submitted)

Frigerio G, Burri M, Bopp D, Baumgartner S, Noll M (1986). Structure of the segmentation gene paired and the Drosophila PRD gene set as part of a gene network. Cell 47:735-746.

Gehring W (1985). The homeo box: A key to the understanding of development. Cell 40:3-5.

Harvey RP, Tabin CJ, Melton DA (1985). Embryonic expression and nuclear localization of Xenopus homeobox (Xhox) gene products. EMBO J 5:1237-1244.

Kessel M, Schulze F, Fibi M, Gruss P (1987). Primary structure and nuclear localization of a murine homeodomain protein. Proc Natl Acad Sci USA 84: in press.

Kleinschmidt JA, Dingwall C, Maier G, Franke WW (1986). Molecular characterization of a karyphilic, histone-binding protein: cDNA cloning, amino acid sequence and expression of nuclear protein N1/N2 of Xenopus laevis. EMBO J 5:3547-3552.

Krumlauf R, Holland PW, McVea JH, Hogan BLM (1987). Developmental and spatial patterns of expression of the mouse homeobox gene, Hox 2.1. Development 99:603-617.

Laughon A, Scott MP (1984). Sequence of a Drosophila segmentation gene: protein structure homology with DNA-binding proteins. Nature 310:25-31.

Lonai P, Arman E, Czosnek H, Ruddle FH, Blatt C (1987). New murine homeoboxes; structure, chromosomal assignment and differential expression in the embryo and in adult erythropoiesis. DNA, in press.

Macdonald PM, Ingham P, Struhl G (1986). Isolation, structure and expression of even-skipped: A second pair-rule gene of Drosophila containing a homeo box. Cell 47:721-734.

Mavilio F, Simeone A, Giampaolo A, Faiella A, Zappavigna V, Acampora D, Poiana G, Russo G, Peschle C, Boncinelli E (1986). Differential and stage-related expression in embryonic tissues of a new human homeobox gene. Nature 324:18-25.

McGinnis W (1985). Homeo box sequences of the Antennapedia class are conserved only in higher Animal Genomes. Cold Spring Harbor Symp Quant Biol 50:263-269.

Mlodzik M, Gehring WJ (1987). Expression of the caudal gene in the germ line of Drosophila: Formation of an RNA and protein gradient during early embryogenesis. Cell 48:465-478.

Müller MM, Carrasco AE, DeRobertis EM (1984). A homeo-box-containing gene expressed during oogenesis in Xenopus. Cell 39:157-162.

Odenwald WF, Taylor CF, Palmer-Hill FJ, Friedrich V, Tani M, Lazzarini RA (1987). Expression of a homeo domain protein in noncontact-inhibited cultured cells and postmitotic neurons. Genes & Development 1:482-496.

Poole SJ, Kauvar LM, Drees B, Kornberg T (1985). The engrailed locus of Drosophila: Structural analysis of an embryonic transcript. Cell 40:37-43.

Regulski M, McGinnis N, Chadwick R, McGinnis W (1987). Developmental and molecular analysis of Deformed; a homeotic gene controlling Drosophila head development. EMBO J 6:767-777.

Scott MP, O'Farrell PH (1986). Spatial programming of gene expression in early Drosophila embryogenesis. Ann Rev Cell Biol 49-80.

Simeone A, Mavilio F, Acampora D, Giampaolo A, Faiella A, Zappavigna V, D'Esposito M, Pannese M, Russo G, Boncinelli E, Peschle C (1987). Two human homeobox genes, c1 and c8: Structure analysis and expression in embryonic development. Proc Natl Acad Sci USA 84:4914-4918.

Schneuwly S, Kuroiwa A, Baumgartner P, Gehring WJ (1986). Structural organization and sequence of the homeotic gene Antennapedia of Drosophila melanogaster. EMBO J 5:733-739.

Cellular Factors in Development and Differentiation:
Embryos, Teratocarcinomas, and Differentiated Tissues, pages 105–123

THE LOCATION, MODIFICATION, AND FUNCTION OF THE *fushi tarazu* PROTEIN DURING *DROSOPHILA* EMBRYOGENESIS

Henry M. Krause and Walter J. Gehring

Department of Cell Biology, Biozentrum, University of Basel, Klingelbergstrasse 70, CH-4056 Basel, Switzerland

SUMMARY

The *fushi tarazu* (*ftz*) protein of *Drosophila* is required during embryogenesis for the process of body segmentation. In order to study the biochemical properties of the *ftz* protein, *ftz* cDNA was expressed in *E.coli*, and the protein purified to homogeneity. Polyclonal antibodies raised against the purified protein were used to localize and quantitate the protein during embryogenesis. Three temporally and spatially distinct phases of expression were observed, which include a previously undetected period later in embryogenesis. During this last phase, the protein is localized in the developing hindgut.

Analysis of embryonic *ftz* protein on Western blots permitted us to approximate the number of protein molecules per nucleus. During the blastoderm phase of development, when *ftz* protein is most abundant, we calculate that there are 15,000 molecules of protein per *ftz*-expressing nucleus.

Since embryonic *ftz* protein migrates more slowly on SDS polyacrylamide gels than protein expressed either in *E.coli*, or *in vitro* in a reticulocyte lysate system, it is apparently modified in the embryo. Two-dimensional gel electrophoresis followed by Western blotting resolves the protein into a series of isoforms which have variable charge and electrophoretic mobility.

When compared in its ability to bind DNA in a sequence-specific manner, it was also found that *ftz* protein partially purified from embryos binds with greater specificity than its bacterially expressed counterpart. In this paper, we demonstate that embryonic _ftz_ protein binds to a specific region within the _ftz_ enhancer element. The potential relationship between these observations is discussed.

INTRODUCTION

The _fushi tarazu_ gene is one of the best characterized genes involved in the control of _Drosophila_ embryological development (for reviews see Gehring and Hiromi, 1986; Scott and O´Farrell, 1986). Embryos which lack _ftz_ protein fail to form alternate body sections, which include the anlagen for approximately half the segmental boundaries (Wakimoto and Kaufman, 1981, Wakimoto et al., 1984). As a result, these embryos end up with about half the normal number of segments, and die before emerging as larvae. The _ftz_ mutation maps to the homeotic gene cluster known as the _Antennapedia_ Complex (Wakimoto and Kaufman, 1981). The gene has been cloned (Kuroiwa et al., 1984; Weiner et al., 1984) and sequenced ((Laughon and Scott, 1984; Kuroiwa, Baumgartner , Krause and Gehring, unpubl. results) and contains a homeo box. The homeobox is a short homologous sequence which is common to a number of developmental control genes (McGinnis et al., 1984a,b; Scott and Weiner, 1984). Within the encoded peptide, the homeodomain, homology has been found to the "helix-turn-helix" motif which enables many regulatory proteins to bind DNA in a sequence-specific manner (Shepherd et al., 1984; Laughon and Scott, 1984). Recent results indicate that the homeotic genes do in fact encode proteins, and that these proteins are nuclearly located (see for example White and Wilcox, 1984; Carroll and Scott, 1985). Furthermore, a few homeodomain-containing proteins have been shown to be capable of binding DNA with some sequence-specificity (Desplan et al., 1985; Fainsod et al, 1986). Recently, Hall and Johnson (1987) have demonstrated that the yeast regulatory protein Matα2, which contains a homeodomain (Shepherd et al., 1984b), binds to specific DNA sequences in a homeodomain-dependent manner.

The transcript of the _ftz_ gene has been identified, and can be detected on Northern blots in abundance between two and four hours of development, and then at reduced levels until about twelve hours of development (Kuroiwa et al., 1984; Weiner et al., 1984). During the first period of expression, the transcripts form seven evenly spaced, zebra-like bands which encircle the embryo perpendicular to the anterior-posterior axis (Hafen et al., 1984). These correspond in location to the regions which are missing in *ftz*⁻ embryos.

Although *ftz* transcripts were not intially detected by *in situ* analysis in embryos older than four hours of development, immunofluorescent staining of *ftz* protein on whole embryos demonstrated the occurrence of a second phase of expression in the developing nervous system (Carroll and Scott, 1985). Patterns of protein located in the nuclei of specific neuronal precursors were found to be repeated in a segmental fashion. Expression of the *ftz* gene in the nervous system was also demonstrated by introducing *ftz* promoter elements, fused to a ß-galactosidase reporter gene, into the embryo.(Hiromi et al., 1985). Recent experiments have demonstrated that *ftz* expression in the nervous system is required in certain neurons for the establishment of proper axonal connections (Doe et al., submitted).

An intact *ftz* coding region flanked by 6.1 kb of 5′ and 1.9 kb of 3′ DNA is sufficient for the rescue of *ftz*⁻ embryos (Hiromi et al., 1985). By fusing the 5′ promoter sequences to a ß-galactosidase reporter gene, three distinct regulatory elements in the 6.1 kb region were defined. Proximal to the transcript initiation site is the "zebra" element which contains the putative binding sites for general transcription factors, and proteins that restrict the expression of the gene to the seven stripe domains observed during the first phase of expression (Hiromi et al., 1985; Hiromi and Gehring, 1987). Adjacent to this and perhaps overlapping, is a region of DNA that is required for expression in the developing nervous system (Hiromi et al., 1985). Approximately 5 kb upstream from the transcription start site is an enhancer element, which also works in both orientations on heterologous promoters (Hiromi et al., 1985; Hiromi and Gehring, 1987). This element was found to be dependent upon the product of the *ftz* gene. Promoter fusions present in *ftz*⁻ embryos were expressed in much the same manner as constructs that were missing the enhancer were in wild type embryos.

Although much is known about the temporal and spatial expression of the *ftz* gene, little is known about the actual properties and roles of the protein. Here, some of those properties are described. Full-length protein expressed in *E.coli* was purified and used to generate antibodies. Immunostaining of embryos using these antibodies confirms the data of Carroll and Scott (1985), and in addition, shows that *ftz* protein is expressed at a third time during a later stage of embryogenesis. It is also shown that *ftz* protein produced in *Drosophila* has a number of properties which are different than protein synthesized in two heterologous systems.

RESULTS

Overexpression of ftz protein in E.coli

In order to investigate the biochemical properties of the *fushi tarazu* protein, we sought to purify the protein to homogeneity so that it might be studied *in vitro*. To facilitate purification, *ftz* cDNA was cloned in a plasmid vector designed to over-express the gene product in *E.coli*. Due to the nature of the cloning scheme, the cloned protein differs from the *Drosophila* protein sequence by only two amino acids added at the amino terminal end.

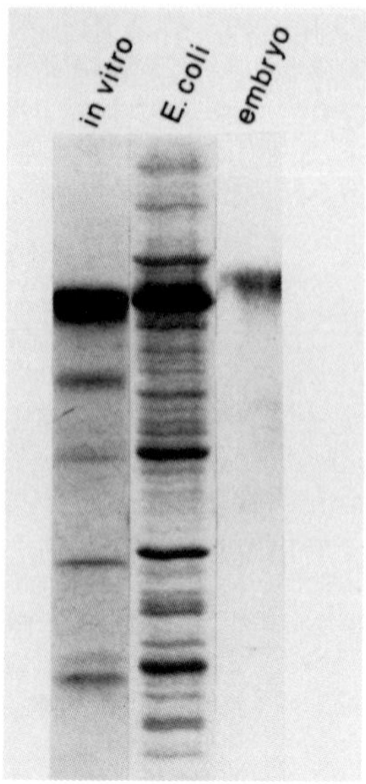

Figure 1. Comparison of protein synthesized in *Drosophila* and in heterologous systems. Samples containing *ftz* protein were resolved on a 12% SDS polyacrylamide gel and represent; 1) (^{35}S)-labeled *ftz* protein transcribed *in vitro* from the pGEMF1 plasmid and translated in a reticulocyte lysate, 2) Coomassie-stained *ftz* protein expressed from the same plasmid in *E.coli*, 3) *ftz* protein from 3-4 hour old embryos detected by immunoblotting.

A number of expression systems were tested. Although all are capable of expressing ftz in vitro (see Fig. 1A), these same systems were not capable of producing detectable levels of protein in bacterial cells. However, when ftz DNA was recloned into a plasmid in which the gene is expressed from a T-7 phage RNA polymerase promoter, ftz protein was made in abundance (Fig. 1B). The protein expressed in E.coli, using the expression system designed by Studier and Moffat (1986) comigrates with the protein synthesized in vitro. Both have an estimated size of 63 kd. Since the predicted molecular weight of the ftz protein is 47.5 kd, and the expression plasmids were found to be free of any cloning errors, we conclude that the reduced mobility of the protein on SDS polyacrylamide gels is a property which is inherent in the protein's primary structure. Indeed, most homeotic and segmentation genes tested also demonstrate this behaviour on SDS polyacrylamide gels (Carroll and Scott, 1985; Ollo and Maniatis, 1987; H. Krause, U. Walldorf, P. LeMotte and B. Dalle Carbonare, unpublished results). Although expression of ftz protein in E.coli using this system reached a cumulative level of 3-5% of total cellular protein, this value is only 10% of what the system is capable of (Stüber and Moffat, 1986). Since we observe a lot of degraded protein on Western blots, proteolysis of the ftz protein may be a major factor limiting its accumulation in bacterial cells.

Having achieved expression of the ftz protein in E.coli, the next step was to develop a purification scheme. We found that lysis of the cells by a variety of techniques and under a variety of conditions yielded insoluble protein. Upon further investigation, it was decided that protein insolubility is a property of the ftz protein when expressed in E.coli, and is apparently not due to the rate of synthesis, the abundance of the protein in the cell, or the method of cell lysis. This conclusion is supported by the observation that ftz protein translated from in vitro-transcribed RNA in a reticulocyte lysate system is also insoluble. Furthermore, we have found that other homeotic and segmentation genes expressed in E.coli using the Studier and Moffat expression system range from being completely soluble to completely insoluble (unpublished results).

A purification scheme was developed, which utilizes urea to solubilize the protein, and DEAE cellulose and dsDNA cellulose column chromatography, to remove other remaining proteins (Krause et al., manuscript in preparation). This one-day purification procedure yields a minimum of 20 mg of pure protein in a single day from one liter of induced cells. The pure protein, still requires the presence of 2 M urea to prevent aggregation and subsequent precipitation.

ftz protein purified in this manner binds DNA sequence-specifically, but tends to aggregate on the DNA substrate. It also binds to a large number of sites, including sequences in the plasmid pBR322 (see below). The problem of protein insolubility taken together with these observations suggests that processes or components of the cells in which the protein is normally expressed may be required to render the protein soluble and to enhance the selectivity of DNA binding. Hence, an attempt to identify and purify the _ftz_ protein made in _Drosophila_ embryos was initiated so that we could compare its properties to that of protein expressed in heterologous systems.

ftz protein made in Drosophila

To initiate this study, antibodies which recognize _ftz_ protein were raised in rabbits against protein purified from _E.coli_. On Western blots, the antibodies recognize the 63 kd _ftz_ protein synthesized in bacteria, and a slightly larger fuzzy band in protein extracts taken from _Drosophila_ embryos (Fig. 1C). This more slowly migrating protein is _ftz_, as will be demonstrated more decisively in a later section of the paper. The ability to detect the endogenous protein demonstrates that these antibodies are specific, and can detect very low levels of protein (0.1 ng can be detected on Western blots). It also permitted the experimental approaches and conclusions which follow in this paper.

One possibility which became immediately obvious was that an approximation of the abundance of protein molecules in a single nucleus could be calculated. The amount of _ftz_ protein shown in figure 1C gives the same signal on Western blots as approximately 1 ng of protein purified from _E.coli_. Since this amount of protein was extracted from approximately 400 embryos, and each embryo contains an estimated 2,000 blastoderm nuclei which express _ftz_ protein (see below), we calculate that the number of molecules per nucleus is approximately equal to 15,000. Although this number seems high in comparison to bacterial regulatory proteins, these nuclei are much larger than bacterial cells. Taking the volume of a blastoderm nucleus into consideration, the concentration of ftz protein works out to be about 50 nM, about two and a half times higher than the concentration of lambda repressor protein in a lambda lysogen.

Since these calculations also indicated that purification of this protein to homogeneity would be extremely difficult (5 kgs or 4×10^8, 3-4 hr old embryos would yield a maximum of 1 mg ftz protein), an overexpression system was designed which could produce higher yields of protein in Drosophila embryos. The ftz gene was fused to the hsp70 heat shock promoter, and integrated into the fly genome by P-element mediated germ line transformation (Krause et al., manuscript in preparation). We found that by heat shocking transformed embryos, ftz protein could be expressed ubiquitously in all nuclei at all times during embryogenesis tested (prior to 2 hrs was not tested). This treatment yields 20 to 50 times more protein per embryo than wild type embryos during their highest period of expression.

ftz protein is modified

It has already been demonstrated that embryonic ftz protein migrates more slowly on SDS polyacrylamide gels than protein synthesized in heterologous systems (see Fig. 1). This suggests that the protein is covalently modified. To test this, we performed two-dimensional gel electrophoresis to see if the Drosophila protein could be further resolved from the protein made in E.coli. Protein samples were separated first by charge using isoelectric focusing gels, and then by molecular weight using SDS polyacrylamide gels (see Fig. 2). We found that the E.coli protein focuses essentially as a single spot with a pK of approximately 7 (Fig. 2A). This is in agreement with the predicted value of 6.8. In contrast, the Drosophila protein synthesized in heat shocked hsp70-ftz transformant embryos is resolved into a series of spots which are more negatively charged than the protein expressed in bacteria (Fig. 2B). A shift in gel mobility is also observed, predominantly with the spots which are closest to the unmodified E.coli protein. Thus, the protein expressed in embryos exists as a series of isoforms which possess variable numbers of covalently bound, negatively charged modifications. These modifications appear to be phosphorylations since they can be removed with

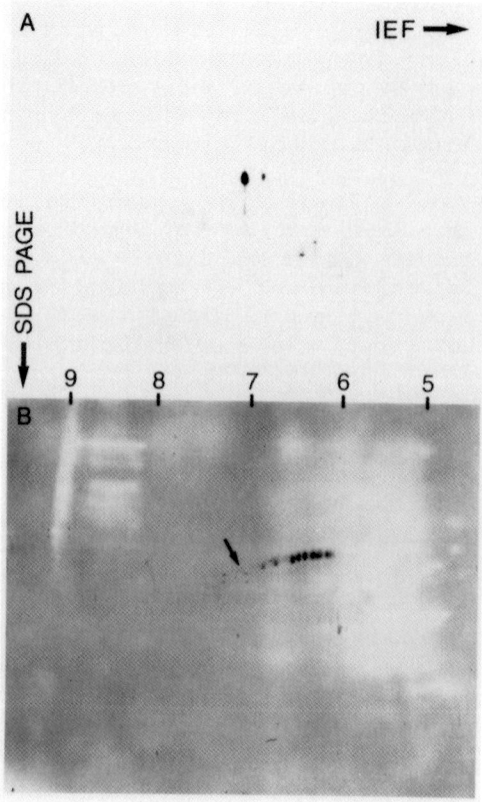

Figure 2. Two-dimensional analysis of _ftz_ protein modifications. Proteins were separated on the first dimension (left to right) by isoelectric focusing, and in the second dimension (top to bottom) by SDS polyacrylamide gel electrophoresis. Proteins were transferred to nitrocellulose, and blots were developed with _ftz_ antibodies, and enzymatic staining of a secondary antibody coupled to horseradish peroxidase. A) _ftz_ protein present in induced bacterial cultures, B) _ftz_ protein expressed ubiquitously throughout _Drosophila_ embryos using embryos transformed with a heat shock-inducible _ftz_ gene. pH´s indicated between the two panels were determined by measuring the pH´s of solutions containing pieces of sectioned isoelectric focusing gels. The arrow in panel B indicates the approximate position at which the major species of _ftz_ protein expressed in _E.coli_ would be located.

phosphatases, and each spot on the 2-D gels becomes labeled when cells or nuclei are incubated with (^{32}P)-orthophosphate (Krause and Gehring, manuscript in preparation). Since the protein can apparently be modified at a minimum of eleven different sites, there is a good chance that some of these phosphorylation events affect particular functional properties of the protein. Interestingly, the *ftz* protein isolated from wild type embryos at different stages of embryogenesis shows very different patterns of phosphorylated isoforms. This suggests that differential phosphorylation during development may be a mechanism by which the function of the *ftz* protein could be modulated in different tissues.

Three phases of expression

The data in Figure 1 demonstrates that antibodies generated against *ftz* protein made in *E.coli* recognize a single species of protein made in wild type *Drosophila* embryos. In order to demonstrate that this protein is synthesized during embryogenesis at the same times expected for *ftz* gene expression, a developmental Western blot was made (see Fig. 3). The blot shows that the protein is first expressed at three hours of development (at 25°C). After a peak in expression between 3 and 4 hrs, the abundance decreases until the 8-10 hr interval whereupon the abundance peaks once again. The amount of protein present in 8-10 hr old embryos is approximately half of that present in 3-4 hr embryos. Interestingly, the message from which the protein is derived is only about 10% of the amount expressed during the initial peak of expression (Kuroiwa et al., 1984; Weiner et al., 1984) suggesting that the protein expressed in the second interval is either more stable, or more efficiently translated. After 12 hours, *ftz* protein could no longer be detected by Western blot analysis.

A second technique was used to determine the spatial localization of the protein during embryogenesis (Fig. 4). *ftz* antibodies were used to detect the protein on whole-mount embryos using immunofluorescence. This shows that during the first phase of expression the protein is localized in seven, zebra-like stripes which encircle the embryo (Fig. 4A-C). In the second phase, the protein is expressed in specific neuronal precursor cells of the developing nervous system (Fig. 4D-G). This is consistent with the results of the developmental Western blot and similar data previously reported (Carroll and Scott, 1985;

Hiromi et al., 1985).

The quality of the antibodies which we prepared also permitted the observation of details which have not been previously reported. For example, figure 4A demonstrates an early intermediate of the first phase of expression. Two details are important. First, the difference in stripe

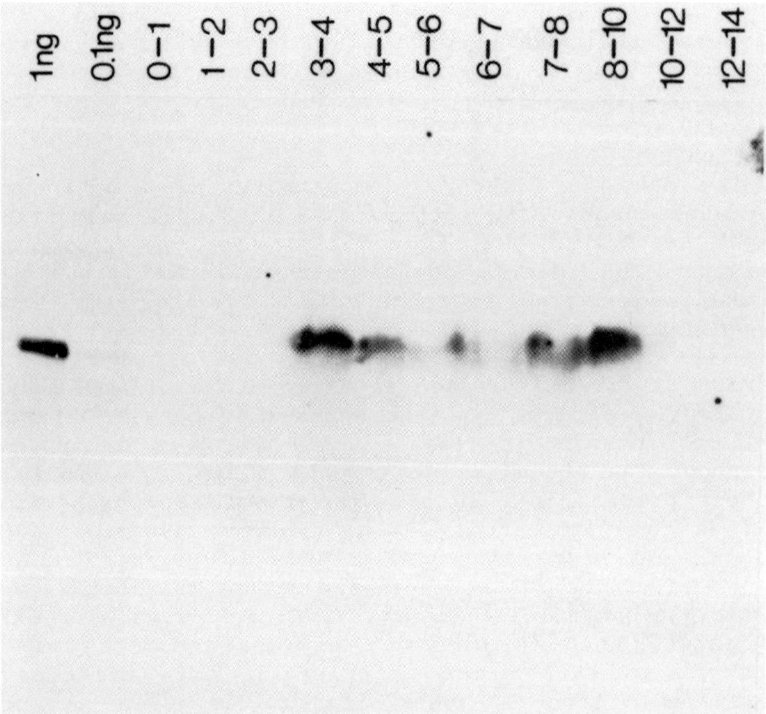

Figure 3. Western blot analysis showing developmental expression of ftz protein.
Embryos collected for one hour (0 through 8 hours) or two hour (8 through 14 hours) intervals were weighed and lysed in 5 µl/mg SDS lysis buffer. 25 µl samples were separated by SDS polyacrylamide gel electrophoresis, and blotted to nitrocellulose. ftz protein was detected using ftz antibodies and (^{125}I)-labeled protein A. Time intervals indicating the ages of collected embryos are shown, as are the amounts of ftz protein purified from E.coli used as size and quantitation standards.

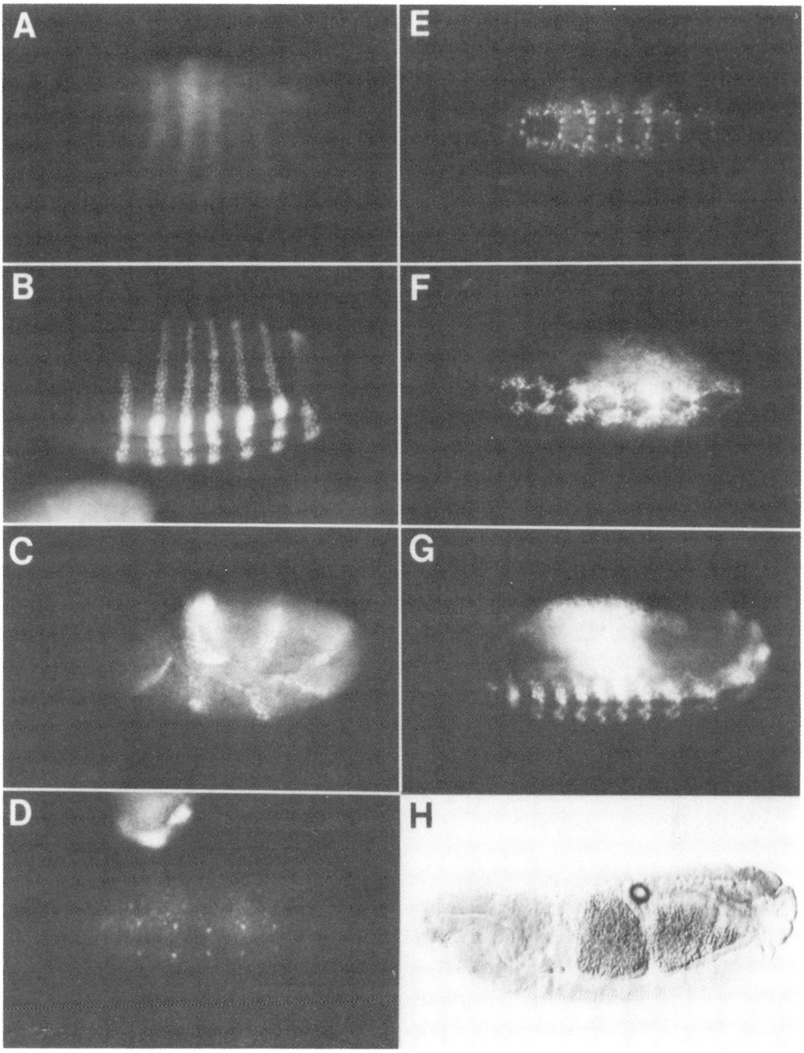

Figure 4. Immunofluorescence localization of ftz protein expressed during embryogenesis
Staged embryos were incubated with anti-ftz antibodies, and a swine, anti-rabbit secondary antibody coupled to fluorescein except in H, where the secondary antibody was coupled to horseradish peroxidase. Embryos are oriented with the anterior end to the left, and ventral surface

down except for the micrographs D-F which are viewed from the ventral side. Based on a 21 hour period of embryogenesis (at 25°C) embryos are estimated to be approximately A) 3 hours, B) 3 1/4 hours, C) 4 hours, D) 6 hours, E) 7 hours, F) 9 hours, G) 10 hours, and H) 14 hours old. Micrographs A-C show stages of ectodermal stripe expression, D-G nervous system expression, and H, hindgut expression. After 9 hours of development, autofluorescence of the yolk proteins present in the developing midgut interferes with the true signal of the secondary antibody. Therefore, the embryo in H was enzymatically stained to detect the secondary HRP-coupled antibody. A dark ring of staining can be seen around a specific sections of the hindgut.

intensities reflects the order of stripe appearance, i.e. first 2, then 1 and 3, 5, 6, 7, and finally stripe number 4 (stripes are numbered from anterior to posterior). Second, the stripes are broad, with poorly defined borders. These become sharply defined in a very short time (less than 15 min). Nuclei at the borders of the stripes either accumulate levels of protein equivalent to those in the middle of the stripes, or when levels are below a certain threshold, expression is shut off. This is consistent with models which suggest that segmentation genes positively regulate their own expression, and repress the transcription of the segmentation genes expressed in adjacent regions (Meinhardt and Gierer, 1980). Later, stripes one through six become narrower, while stripe seven becomes broader. These changes appear to occur predominantly on the posterior sides of the stripes. Recently, this has also been observed by Lawrence et al. (1987). Since the anterior edges of _ftz_ stripes correspond to those of overlapping _engrailed_ stripes, they suggest that the major role of early _ftz_ expression is to delimit parasegmental boundaries.

Our antibodies also detected a third phase of _ftz_ expression (Fig. 4H). This phase occurs between twelve and fifteen hours of development and is predominantly localized to a section of the developing hindgut just posterior to the site of Malpighian tubule attachment. Due to the shape of the gut, this appears as a tube or ring, depending on the angle at which it is viewed. This expression is not abundant enough to be detected on Western blots.

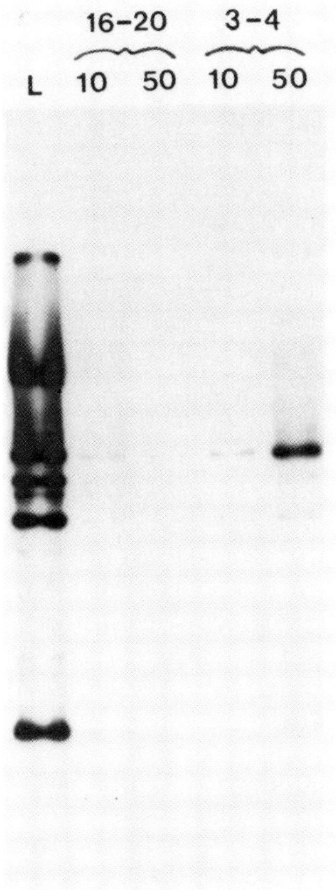

Figure 5. Binding of ftz protein isolated from Drosophila to the ftz enhancer. Partially purified protein isolated from Drosophila embryos was incubated with (^{32}P) end-labeled restriction fragments. Binding of partially purified Drosophila ftz DNA is shown by immuno precipitation of fragments to which ftz protein is bound. Labeled DNA was separated by polyacrylamide gel electrophoresis and visualized by autoradiography. DNA was incubated together with two different concentrations of protein isolated from either 16-20 or 3-4 hr old embryos. The fragment bound is a 1.2 kb EcoR1 fragment located within the ftz enhancer.

Sequence-specific DNA binding

Both ftz protein purified from bacteria, and partially purified from Drosophila embryos, have been used in DNA binding experiments. Both bind DNA in a sequence-specific manner, but with apparent differences. As mentioned previously, the bacterially-expressed protein tends to aggregate on the DNA in a seemingly cooperative process. This is supported by several observations. 1) When non-denaturing polyacrylamide gels are used to visualize DNA restriction fragments which are shifted in mobility by specifically-bound protein, the bands are not shifted to defined positions. Rather they become shifted to the top of the gel within a very narrow range of ftz protein concentrations. 2) Visualization of protein-DNA complexes by electron microscopy shows that the protein binds first to defined locations which are consistent with other binding assays (H. Krause and C. Brack, unpubl. results). However, these complexes are huge, and at slightly higher concentrations of _ftz_ protein, complex networks of protein and DNA are seen. A few of these have very ordered helical structures of protein which cover the whole DNA molecule. 3) DNase I footprinting experiments show defined regions of protection on specifically bound fragments, however, these are poorly protected and have ill-defined borders. As soon as more protein is added, the whole fragment is protected from DNase 1 digestion, and in fact, is shifted upwards in the 8 M urea-containing sequencing gels.

In contrast, protein extracted from _Drosophila_ embryos shifts bands in gel retardation assays to discrete locations. Also, defined regions of protection are seen in DNase 1 protection experiments (Krause and Gehring, manuscript in preparation). The _Drosophila_ protein binds to the majority of DNA fragments bound by the bacterially-expressed protein, but does not bind to pBR322 plasmid sequences (see for example, Fig. 5). Finally, the protein extracted from embryos also appears to have different relative affinities for sites which are bound by both proteins.

Binding of _ftz_ protein to DNA sequences from a number of genes has been investigated. Binding sites have been identified in the vicinity of the homeotic genes _Antennapedia_ (_Antp_) and _Ultrabithorax_ (_Ubx_) and the segmentation gene _engrailed_, as well as the _ftz_ gene itself. All of these genes have been shown to be expressed in a _ftz_-dependent manner (Duncan, 1986; Ingham and Martinez-Arias, 1987; Howard and Ingham, 1986; Hiromi et al., 1985). Figure 5 shows sequence-specific binding of _ftz_ protein extracted from wild type _Drosophila_ embryos to plasmid DNA containing _ftz_ upstream promoter sequences. In this experiment, DNA that _ftz_ protein specifically interacts with was isolated from the binding reaction by immunoprecipitation and detected on a polyacrylamide gel. Binding only occurs to a restriction fragment which is derived from the _ftz_ enhancer element. As a control, we demonstrate that only protein from 3-4 hr old embryos, and not from 16-20 hr old embryos can achieve this result. Thus, _ftz_ protein binds to its own enhancer. We have also found binding sites in the zebra element, and 3′ to the coding region, however, the enhancer site is the highest affinity site which we have identified.

DISCUSSION

A number of very important differences have been demonstrated between the biochemical properties of ftz protein made in Drosophila embryos, and in heterologous systems. These include protein solubility, post-translational modifications and DNA sequence specificity.

Protein which was partially purified from Drosophila embryos did not have a tendency to precipitate out of solution as it did when made in heterologous systems. Although this property may be linked to protein modifications, there are many factors present in these preparations which may also be responsible for this property.

We have shown that ftz protein purified either from embryos or bacteria, binds DNA in a sequence-specific manner. The Drosophila protein, however, binds to fewer sequences, and with different relative affinities than the E.coli-made protein. The stoichiometry of binding was also found to be different. Specific binding sites on the DNA appear to act as focal points for large-scale aggregation of protein purified from E.coli. This seemingly cooperative binding was not observed with protein isolated from embryos.

Binding sites have been identified in the vicinity of a number of homeotic and segmentation genes. In the case of the ftz gene, regulatory elements have been identified (Hiromi et al., 1985) and we find that the strongest ftz protein binding site occurs in the ftz enhancer element. Since it has been demonstrated that the ftz enhancer element requires the ftz protein itself for quantitative, and sustained expression (Hiromi et al., 1985; Hiromi and Gehring, 1987) this suggests that the effect is mediated through the ability of the protein to interact site-specifically with its own enhancer. It is possible that ftz and other homeodomain-containing proteins may also activate the expression of other genes through their ability to bind and activate enhancer elements associated with those genes. It remains to be demonstrated whether or not binding of ftz protein to its own enhancer actually affects transcription of the ftz gene.

We have shown that during development, the ftz protein is expressed in three very different tissues during three phases of embryogenesis. In the first phase the protein is expressed in alternate segment primordia in a pair-rule fashion. In the second, the protein is localized in a segmentally repeating pattern in specific neuronal precursor cells and neurons of the developing nervous system. Finally, the protein is expressed in the gut in a non-segmental pattern. The protein is nuclearly located in all cells.

During the first phase of expression, the protein is required for the formation of segmental boundaries. In the nervous system, _ftz_ expression appears to control the lineage of defined neurons (Doe et al., 1987). No function has yet been associated with its expression in the hindgut. Although we know a little about what the protein is required for during the first two phases of expression, we still have no idea how these effects are achieved. The processes of cell differentiation which are controlled may be effected at many different levels. The elaboration of these processes will probably require the identification of downstream genes which are controlled by _ftz_ in the different sites of expression. Some of these will probably be tissue-specific, and only turned on during one or another phase of _ftz_ expression.

When the _Drosophila_ protein was examined, it was found that the protein is phosphorylated at different sites on the protein and that different subsets of isoforms are made during different phases of expression. This may be the first example of a gene regulatory protein which is differentially phosphorylated during development. The potential of differential phosphorylation to generate a number of protein isoforms with unique biochemical properties may explain how the same protein could be utilized by different cells to control different regulatory events. Various tissue specific protein kinases and phosphatases could then generate a diversity in regulatory potential from a minimal number of master control genes. Protein phosphorylation, and post-translational modifications in general are obviously important processes that are utilized by eukaryotic cells to regulate cellular processes. The results reported here suggest that they may also play a role in early development through the modulation of proteins which regulate processes of pattern formation.

REFERENCES

Carroll SB, Scott MP (1985). Localization of the fushi tarazu protein during Drosophila embryogenesis. Cell 43:47-57.

Doe C, Hiromi Y, Gehring WJ, Goodman G (1987). Specific neuronal precursor cells and progeny express the Drosophila segmentation gene fushi tarazu. manuscript submitted.

Desplan C, Theis J, O'Farrell PH (1985). The Drosophila developmental gene, engrailed, encodes a sequence-specific DNA binding activity. Nature 318:630-635.

Duncan I (1986). Control of Bithorax Complex functions by the segmentation gene fushi tarazu of D. melanogaster. Cell 47:297-309.

Fainsod A, Bogard LD, Ruusala T, Lubin M, Crothers DM, Ruddle FH (1986). The homeodomain of a murine protein binds 5' to its own homeo box. Proc. Natl. Acad. Sci. USA 83:9532-9536.

Gehring WJ, Hiromi Y (1986). Homeotic genes and the homeobox. Ann. Rev. Genet. 20:147-173.

Hafen E, Kuroiwa A,Gehring WJ (1984). Spatial distribution of transcripts from the segmentation gene fushi tarazu during Drosophila embryonic development. Cell 37:833-841.

Hall M, Johnson A (1987). Homeodomain of the yeast repressor Matα2 is a sequence-specific DNA binding domain but is not sufficient for repression. Science 237:1007-1011.

Hiromi Y, Kuroiwa, A, Gehring WJ (1985). Control elements of the Drosophila segmentation gene fushi tarazu. Cell 43:603-613.

Hiromi Y, Gehring WJ(1987). Regulation and function of the Drosophila segmentation gene fushi tarazu. CELL in press.

Howard K, Ingham P (1986). Regulatory interactions between the segmentation genes fushi tarazu, hairy, and engrailed in the Drosophila blastoderm. Cell 47:949-957.

Ingham PW, Martinez-Arias A (1986). The correct activation of Antennapedia and Bithorax Complex genes requires the fushi tarazu gene. Nature 324:592-597.

Krause H, Klemenz R, Gehring WJ (1987). Comparison of ftz protein made in Drosophila and in heterologous systems: the Drosophila protein is expressed during three phases of embryogenesis and is modified. manuscript in preparation.

Krause HM, Gehring WJ (1987A). The fushi tarazu protein is differentially phosphorylated at different stages of expression during Drosophila embryogenesis. manuscript in preparation.

Krause HM Gehring WJ (1987B). fushi tarazu protein partially purified from Drosophila embryos binds DNA in a sequence-specific manner. manuscript in preparation.

Kuroiwa A, Hafen E, Gehring WJ (1984). Cloning and transcriptional analysis of the segmentation gene fushi tarazu of Drosophila. Cell 37:825-831.

Laughon A, Scott MP (1984). Sequence of a Drosophila segmentation gene: protein structure homology with DNA-binding proteins. Nature 310:25-31.

Lawrence P, Johnston P, Macdonald P, Struhl G (1987). Borders of parasegments in Drosophila embryos are delimited by the fushi tarazu and even-skipped genes. Nature 328:440-442.

McGinnis W, Levine MS, Hafen E, Kuroiwa A, Gehring, WJ (1984a). A conserved DNA sequence in homoeotic genes of the Drosophila Antennapedia and Bithorax Complexes. Nature 308:428-433.

McGinnis W, Garber RL, Wirz J, Kuroiwa A, Gehring WJ (1984b). A homologous protein coding sequence in Drosophila homeotic genes and its conservation in other metazoans. Cell 38:403-409.

Meinhardt H, Gierer A (1980). Generation and regeneration of sequence of structures during morphogenesis. J. Theor. Biol. 85:429-450.

Scott MP, O´Farrell PH (1986). Spatial programming of gene expression in early Drosophila embryogenesis. Ann. Rev.

Cell Biol. 2:49-80.

Scott MP, Weiner AJ (1984). Structural relationships among genes that control development: Sequence homology between the *Antennapedia*, *Ultrabithorax*, and *fushi tarazu* loci of *Drosophila*. Proc. Natl. Acad. Sci. USA 81:4115-4119.

Shepherd JCW, McGinnis W, Carrasco AE, De Robertis EM, Gehring WJ (1984). Fly and frog homeo domains show homology with yeast mating-type regulatory proteins. Nature 310:70-71.

Studier WF, Moffatt BA (1986). Use of T-7 RNA polymerase to direct selective, high-level expression of cloned genes. J. Mol. Biol. 189:113-130.

Wakimoto BT, Kaufman TC (1981). Analysis of larval segmentation in lethal genotypes associated with the *Antennapedia* gene complex in *Drosophila melanogaster*. Develop. Biol. 81:51-64.

Wakimoto BT, Turner FR, Kaufman TC (1984). Defects in embryogenesis in mutants associated with the *Antennapedia* gene complex of *Drosophila melanogaster*. Develop. Biol. 102:147-172.

Weiner AJ, Scott MP, Kaufman TC (1984). A molecular analysis of *fushi tarazu*, a gene in *Drosophila melanogaster* that encodes a product affecting embryonic segment number and cell fate. Cell 37:843-851.

White RAH, Wilcox M (1984). Protein products of the *Bithorax Complex* in *Drosophila*. Cell 39:163-171.

Cellular Factors in Development and Differentiation:
Embryos, Teratocarcinomas, and Differentiated Tissues, pages 125–145

RECOMBINANT RETROVIRUS, CELL LINEAGE AND GENE EXPRESSION IN THE MOUSE EMBRYO

Jean-François Nicolas and Claire Bonnerot

Unité de Génétique Cellulaire du Collège de France et de l'Institut Pasteur,25 rue du Docteur Roux,75015 Paris, France

INTRODUCTION

The basis of tissue and temporal expression of genes is believed to reside in a combination of a number of discrete *cis*-acting elements generally grouped in the proximity of the promotors (reviewed in Maniatis et al., 1987; Palmiter and Brinster, 1986). This combination of modulators is recognized by *trans*-acting factors which regulate transcription positively or negatively (reviewed in McKnight and Tijan, 1986) ; the expression of these modulators, (and their secondary modifications) must be developmentally controlled. The acquisition of the specific pattern of *trans* -acting elements as well as of other characteristics of differentiated cells is the result of successive choices which have progressively determined their nature. It is generally assumed that a series of binary decisions underly cell specification (Kauffman, 1975) and that each choice is made in a strict relation with either the lineage of the cell, its position or local cell interaction. Examples of each type of process have been described in those organisms for which cell lineage and genetic studies are easy to perform (reviewed in Gardner and Lawrence, 1985).

In mammals the existing information on cell lineage, its relevance to the spatial and temporal expression of genes and morphogenesis is very incomplete. It is assumed that development is much more labile, that cell mixing is prevalent, and that cellular interactions are predominant to determine cell fate. This picture, however is far from being firmly established and in part it may reflect the imprecisions of our present knowledge of these problems, due to not only the complexity of the embryo and its inaccessibility to experimental manipulations, but also to the lack of a general method to perform clonal analysis (reviewed in Rossant, 1987).

The properties of defective recombinant retroviruses (reviewed in Nicolas and Rubenstein, 1987) - as a way to introduce *in vivo* genetic material into cells - and of an *in situ* detectable indicator gene have been combined to provide a technique for single cell marking and creation of induced genetic mosaicism during embryogenesis (Sanes et al., 1986 ; Bonnerot et al.,1987). In addition, the availability of in situ markers has made it possible to define precisely the spatial and temporal expression of several promoters during the first stages of embryogenesis (Goring et al., 1987 ; Bonnerot et al.,1987).

In this article we will discuss the various problems related to these new approaches. In particular, (1) the production of stocks of defective recombinant retrovirus, (2) the properties of the *in situ* markers, (3) the way to express the marker with minimal interference from the viral elements, and (4) its applications to the studies on cell lineage in post implantation embryos and gene expression during early mouse embryogenesis.

RETROVIRAL VECTORS.

Construction and Production.

The property of retrovirus which must be conserved in vectors lies in their ability to produce an RNA which possesses the two main features of viral genomic RNA : their ability to be packaged in the viral particle, and to be reverse transcribed into a DNA copy which can integrate into the host genome as a defective provirus.

The strategy adopted to produce such a defective recombinant retrovirus (dRRV) is presented in Fig. 1 (see also Fig. 2D for the structure of the long terminal repeat and Coffin,1985 ; Nicolas and Rubenstein, 1987 for additional informations). There are two distinct elements to engineer. The first one is a DNA molecule (pRRV,Fig.1) which will produce a genomic RNA with a chimeric structure (cgRNA). The plasmid pRRV includes : (1) a complete long terminal repeat (LTR) and about 200 adjacent bases, positioned 5' to the insert. This fragment provides a promoter (for the transcription of the cgRNA starting at the first base of the R region), the R region (involved in the first jump of the DNA small strong stop during reverse transcription), the binding site for the tRNA primer for minus strand DNA, the Ψ signal (for the packaging of the genomic RNA) and sequences for proviral integration, (2) the insert, (3) a second LTR positioned 3' to the insert which provides a polyadenylation region (in U3), the R region and elements for the plus strand synthesis and integration. The complete integrity of the 3' U3 region is not required (see below). The second element is a packaging cell line to provide, in

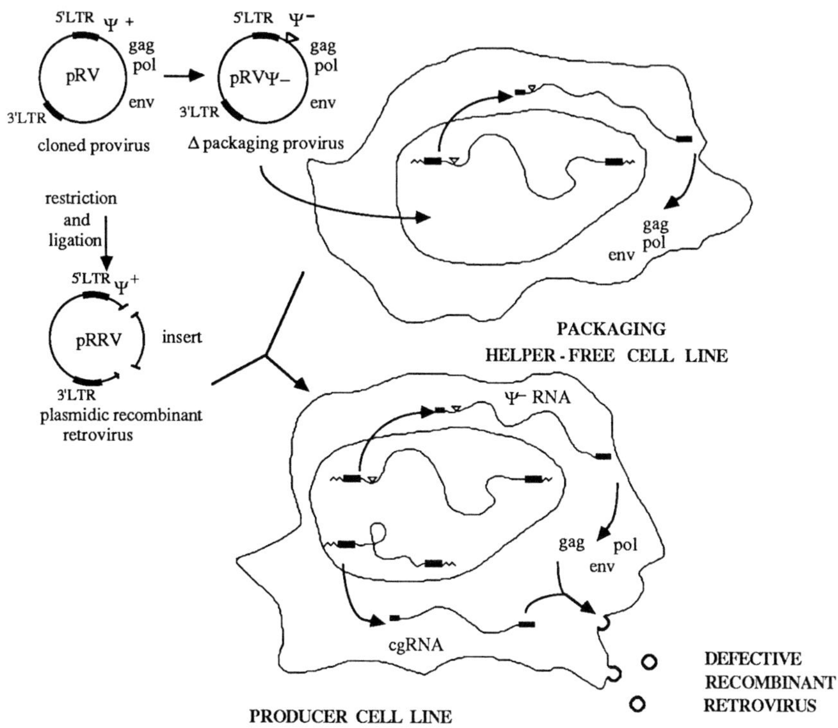

Figure 1 : Construction and production of dRRV.

pRRV which contains the elements required in *cis* to produce a dRRV is derived from a cloned wild type provirus (pRV) like pMov3 (Jähner and Jaenisch, 1980). The gag, pol and env sequences of the provirus can be deleted, although the presence of 5' sequences of the gag gene may increase the titer of the dRRV (Miller and Buttimore, 1986). The Ψ signal, the 5'LTR and part of the 3'LTR must also be conserved (see text).The upper limit of the cgRNA is around 9 kb.

The producer cell line is derived from the packaging helper free cell line by transfecting it with pRRV. If pRRV has a complete 3'LTR, the producer cell line can also be produced by infecting the ecotropic packaging cell line with an amphotropic virus or vice versa. The packaging helper free cell line is, in general, a mouse 3T3 cell line expressing gag pol and env from Ψ^- copies of an N-B tropic (Risser and Horowitz, 1983), ecotropic or amphotropic provirus introduced into the cells by transfection. Attempts are in progress to obtain the expression of gag pol and env (in the proper ratio) from two complementary Ψ^- molecules deleted for the 3'LTR in order to decrease the production of recombinant wild type virus (see text).

trans, the factors for the assembly and maturation of the viral particles. It could be a M-MuLV infected cell line or, better, an engineered helper-free cell line which constituvely express the viral genes from a unique (Fig. 1) or several non packageable mRNA (Mann et al., 1983 ; Danos et Mulligan, personal communication).

If the pRRV DNA is transcribed into the packaging cell line, recombinant retroviral particules are produced by budding from the cell surface. They can then be harvested directly from the culture medium. From transcomplementing cell lines which constitutively express the viral genes from a Ψ^- provirus, the probability to produce wild type viruses by recombination between the vector and the packaging system is not negligeable (Mann and Baltimore, 1985; Miller and Buttimore, 1986). In addition, low level transmission of the deleted viral genome can also occur. For these reasons, in the future, the transcomplementing cell lines which express the viral genome from two distinct Ψ^- molecules, one coding for gag-pol, the other for env, both deleted from the 3' LTR will replace the ones actually being used. It will decrease further the probability of transfer of functional gag pol and env genes.

Structure of the Inserted Material.

The insert may contain only the sequences coding for the structural part of a gene. If its expression is required, it must be positioned in such a way that it utilizes, as a promotor, the 5' LTR (see pM-MuLVneo in Fig.2). In these kind of vectors, the insert acquires the transcriptional specificity of the viral LTR remaining repressed in multipotential embryonic cells. In order to modify the range of expression of the insert two other modes of transcription have been analysed : (1) by adding a second promotor in an internal position in the provirus (Rubenstein et al., 1984 ; Wagner et al., 1985) and (2) by deleting part of the U3 region of the LTR (Yu et al., 1986).

dRRV with Internal Promotors. These kind of constructs (see pM-MuLVSVtkneo in Fig.2,B) create an unusual structure since two promoters drive the expression of a unique coding sequence. It is, *a priori,* difficult to predict which one will function in the vector and the provirus, since it depends on the relative strength of the LTR compared to the second transcriptional region, a factor that varies depending on the cell line. In addition, trancriptional modulators frequently act at a distance and hence the two promotors may influenced each other. From our experiments the promotors from the SV40 early region, from the HPRT gene or the SVtk promotor can all be inserted as an additional internal promotor with no detectable effect on dRRV production. The fact that

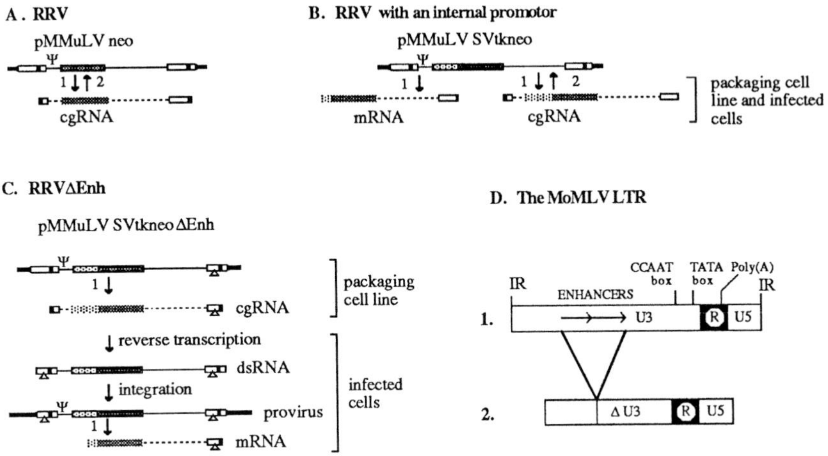

Figure 2 : Different families of dRRV.
Signification of the symbols

- the LTR with (from 5' to 3') U3,R and U5
- the internal promotor
- the inserted gene
- flanking cellular DNA
- retroviral sequences

In wavy lines, the structure of the RNA is drawn.

↓ 1 shows the transcription from the proviral copy

↑ 2 shows the reverse transcription if the cell provides the necessary trans elements

A. dRRV with a unique insert containing no regulatory elements. Only one RNA species is produced.
B. dRRV with an insert driven by an internal promotor. It produces two RNA species with different structures, one of which may be reverse transcribed. The relative quantity of mRNA compared to cgRNA depends on numerous factors (see text).
C. dRRV in which a 3'U3 has been deleted for the two 85 bp repeats. The deletion is transferred to the 5' LTR in the infected cells. The packaging cell line produced two RNA species. One has a cgRNA structure. The infected cells produced almost only mRNA.
D. General features of MoMLV LTR (1) and of a LTR deleted for the two 85 bp long repeats (2). IR : the inverted repeat required for integration ; they delineate the border of the LTR. For details see text.

other internal promotors may have severe effects on virus production is demonstrated by the promotor of the ß-actin gene. In helper cell lines transfected with pMMuLVßact neo, the RNA is initiated almost exclusively from the promotor of ß actin (Bonnerot and Nicolas, unpublished).

In the proviral state, the internal promotor is used either in conjonction with the LTR (in the majority of differentiated cells) or alone (when the 5' LTR is repressed, as in embryonal carcinoma cells).

Alterations in the LTRs. Because in the majority of differentiated cells the LTR is used in addition or in place of the internal promotor, attempts have been made to delete parts of the promotor contained in the LTR (Fig. 2, D). This is possible because although the presence of a functional 5' LTR is required in the producer cell line (for cgRNA production), it is dispensible in the provirus. The pecularity of proviruses is that they are DNA copies of an RNA. Because the proviruses must also be transcribed, retroviruses have developped a way to create a new promotor after each cycle of reverse transcription of the genomic RNA. It is accomplished by two successive jumps of intermediate DNA molecules from one extremity of the gRNA to the other (Gilboa et al.,1979). As a consequence the 5' promotor (U3) in a provirus is derived from the 3' LTR of the preceeding provirus (Fig. 2, C). Several groups have taken advantage of this feature to construct dRRV whose cgRNA originates from a transfected construct harboring a modified 3' U3 (generally a deletion of the two 85 bp repeats, see fig 2 D, or of the whole promotor) but a complete 5' LTR and are converted into a provirus harboring the modification in both LTRs. We have developed such a dRRVΔEnh from vectors with a variety of internal promotors. In all infected cells (differentiated or not) the internal promotor is now mainly used, as judged by the analysis of the structure of the transcribed RNAs (Bonnerot and Nicolas, unpublished).

In conclusion, three families of defective RRV which have distinct transcriptional properties are currently being used (1) dRRV in which the inserted material uses the regulatory elements of the retrovirus, (2) dRRV in which the insert contains its own transcriptional region and (3) dRRV with an internal promotor and altered LTRs. The choice of the dRRV to be used depends essentially on considerations on the nature of the cell in which the transduced genes must be expressed. When expression is only desired in cultured cells the activity of the LTR or internal promotors can easily be pre-checked and if they are active (a common situation) then, dRRV of group A or B (Fig. 2) can be choosen. Their advantages over dRRVΔEnh is that they can be produced at higher titers and are rarely subjected to rearrangements. The dRRVΔEnh are more adapted when the targeting of gene expression is desired and for almost all *in vivo* situations. This is because, if the cells infected with a dRRV of class A and B produce an RNA which can be reversed transcribed, (↑2, in Fig. 2), those infected with dRRV ΔEnh do not, since they use only the internal promotor (Fig.2 C). Thus it decreases the probability of secondary infections from the initially infected cells. These constructs are steps in the conception of a vector whose elements for its own transcription, transport and integration would not interfere with the expression of the insert in the provirus.

CELL MARKERS

Once introduced into a cell, dRRV can be detected by the presence of a provirus in the host DNA and by their expression. For *in vivo* clonal analysis, it is not easy to use the proviral DNA as a marker because of the actual difficulty to detect a single gene copy by *in situ* hybridization techniques (Lo, 1986) especially for preselection of the positive portions of the embryo. In contrast, the expression of an introduced gene can be used as a cell marker. It is advantageous to use a gene encoding an enzyme since this allows an amplification of the signal by its catalytic activity.

The ideal cell marker would fulfill a number of criteria which can essentially be grouped into two categories : those depending only on the properties of the marker (it must be be cell localized, cell autonomous, easy to detect and developmentally neutral) ; those depending in addition on the system used to introduce and express the marker. It includes stability and ubiquity of expression throughout development.

We (Sanes et al., 1986 ; Bonnerot et al., 1987) and others (Price et al.,1987) have analysed whether the ß-galactosidase of E. Coli could be used as such a marker. The bacterial ß-galactosidase can be detected in mammalian cells either by its enzymatic activity or by using specific antisera. With both methods the labelling is sufficiently precise to attribute the expression to a particular cell (see Fig 3) and the level of the background due to galactosidase-like activity is generally not a problem especially in cultured cells. Nevertheless for a number of *in vivo* situations, a strictly non ambiguous marker in term of cell localization and background activity (Dewey and Mintz, 1978) is required. The development of a ß-galactosidase marker that is targeted towards the nucleus would help in solving these difficulties. Indeed when ß-galactosidase is fused to the nuclear locating signal (nls) of the large T antigen of SV40 by its amino terminal (Kalderon et al., 1984) it retains the enzymatic activity and is directed towards the nucleus (Bonnerot et

interfere with the internal promotors in a somehow unexpected way. Thus, precise chararacterization of the constructs in multipotential cells and pre-implantation embryos must precede their utilization.

Multipotential Cells in Culture

Because of the lack (until recently) of any promotor from a gene whose expression is ubiquitous, efforts have been concentrated on the study of viral promotors (Nicolas and Berg,1983 ; Linney et al., 1984 ; Herbomel et al.,1984 ; Gorman et al.,1985 ; Sassone-Corsi et al.,1985 ; Taketo at al., 1985). The picture emerging from a number of independant studies is shown in Table 1 where the results of the expression (in transient assays) of LacZ-based vectors in two multipotential embryonal carcinoma cell lines, PCC3 and PCC4 is reported (Rocancourt, Bonnerot, Nicolas, unpublished). Because in all constructs LacZ expression does not require any splicing event and because, the polyadenylation signal is derived from either SV40 or M-MuLV (known to be utilized in EC cells), we consider it likely that the expression of LacZ essentially reflects the activity of the promotor. The LTR of M-MuLV (pMMuLVnlsLacZ) which gave less than 0.1% ß-galactosidase positive EC cells compared to fibroblastic cells (3T3, WOP, Dailey and Basilico,1985) is extremely inefficient in promoting LacZ expression. In contrast the 342bp long fragment of SV40 (containing the whole early promotor) in pCH110 (Hall et al.,1983) and L7RHß-gal (Kalderon et al.,1984) is very efficient (25% to 100% of ß-gal positive cells compared to 3T3, WOP), as is the LTR of the Rous Sarcoma virus in RSVß-gal (Gorman et al., 1982).

To obtain a greater insight into the way in which promotors are recognized in EC cells, the SV40 early region has been further dissected and studied in transient expression assays using pSV2gpt derivatives (Fig. 3, A, Nicolas, Kadesh, Rocancourt, unpublished). To abolish the contribution of potential promotor sequences in bacterial DNA, two polyadenylation signals have been introduced upstream of the gpt gene. When the SV40 early promotor includes the two 72bp repeats (in pSV2Agpt), gpt activity is easily detected in the four EC cell lines studied. When they are deleted (in pSV232Agpt), the activity is barely detectable in either EC or diffentiated cells. Their reintroduction downstream from the gene (in pSV232Agpt LR) is associated with the reappearence of more than 50% of the initial activity. In addition, when the minimal promotor of SV40 is exchanged for the ßglobin (in itself very poorly active) in either pSV2Agpt (pSßgpt) or pSV232A gpt LR (pSßgpt LR) it is also enhanced in both cell types. Thus, the 72bp repeats are clearly active in multipotential cells (at least in the transient assays) as enhancers acting in *cis* at a distance and active on heterologous promotors. It has been further documented in constructs in which the whole SV40 promotor is

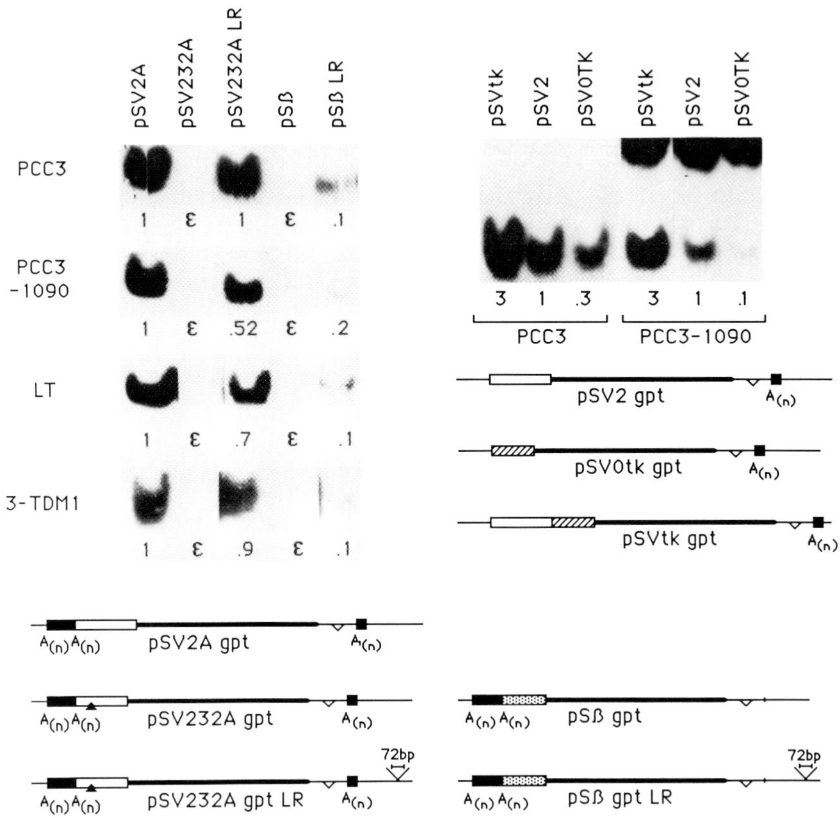

Figure 4 : Enhancer activity of the SV40 72 bp element in EC cells. Expression of the gpt gene introduced as supercoiled DNA in EC cells (LT, PCC3 and PCC3-1090, a nullipotent subclone) and differentiated cells (3-TDM1, Nicolas et al.,1976) has been compared by a gel-transfer assay (Mulligan and Berg,1980) and quantified (the numbers under the spots, ε means barely detectable). The structure of the plasmids is illustrated under the figures.

- polyadenylation signal
- SV40 early promotor
- SV40 early promotor deleted for the 72 bp repeats
- promotor of the Herpes thymidine kinase gene (Nicolas and Berg, 1983)
- promotor of the ßglobin gene
- gpt gene
- bacterial sequences

For maximum expression in EC cells the 72 bp element is required. It acts as a typical enhancer element.

positioned 5' from the tk promoter (Fig. 3,B). These experiments suggest that *trans*-acting factor(s) interacting with one or several of the sub-elements of the enhancer region of SV40 are present in multipotential cells. It should be noted that the SVtk promotor region gives more gpt activity than SV40 alone, probably due to some synergistic effect of the two promotors.

The results obtained in transient assays can essentially be duplicated in experiments on stable transformants obtained by transfection or retroviral infection. Even after integration into the host genome, the LTR of M-MuLV remains inactive but the SV40, SVtk and RSV promotors are potent (Nicolas and Berg, 1983 ; Rubenstein et al., 1984).

However the fine dissection of the SV40 and M-MuLV promotors and their study in differentiated cells have revealed an extremely complex arrangement of regulatory sequences which, individually, can be recognized specifically by different mature cells (Davidson et al., 1986 ; Sen et Baltimore, 1986 ; Schirm et al., 1987). This combination of sequences has been selected by the virus to perform very specific functions. In addition, results of studies on the expression of the SV40 early regions in transgenic animals have shown that their pattern of expression is restricted to a very small number of cell types (Van Dyke et al., 1985). Thus a lot more must be learned before being able to manipulate these promotors at will .

For these reasons we have turned our attention to promotors of genes whose expression is ubiquitous. In studies to be reported elsewhere (summarized in Table 1 for the LacZ constructs) we demonstrated that the promotor of the ß-actin (Nudel et al.,1983) and HPRT (Friedman, unpublished) genes are used in neo and LacZ chimeric construtcs at levels similar to the SV40 early promotor in EC cells. Further studies will show whether they can be used in dRRV for clonal analysis.

Taken together these studies also shed light on the causes for the resistance to infection of multipotential cells (Kelly and Condamine,1982). For M-MuLV it resides in the promotor, but for SV40 it resides in another element, maybe alternative splicing as previously suggested (Nicolas and Berg,1983). In addition they have also provide efficient vectors to introduce by cotransfection any genes into EC or EK lines.

1 and 2 Cell Embryo

The state of differentiation in the normal development to which different embryonal carcinoma and EK cell lines correspond is not a unique one (Nicolas et al.,1981;Jakob and Nicolas,1987). For example some female EC cells may have only one of their X chromosomes active

Vector	ref	Promotors (size)	Expression in PCC3	PCC4	2-cell stage
1) Control					
pCH110ΔSV	1	none	-	-	-
2) Viral promotors					
pCH110	2	SV40e (342 bp)	1,0 - 0,5	0,2 - 0,1	nd
L7RH ßgal	3	SV40e (342 bp)	0,25 - 0,15	0,2	+
pMMuLVnlsLacZ	1	MoMLV (448 bp)	0,0002	0,0006	-
pRSVßgal	1	RSV (230 bp)	yes	yes	+
pMMuLVSVLacZ	4	MoMLV + SV40e	0,02	nd	+
pA56nlsLacZ	7	SV40e (21bp)	-	-	+
3) Genes promotors					
pMMuLVßact LacZ	1	MoMLV + ßact (2 kb)	1,0 - 0,5	nd	+
pMMuLVHPRTnlsLacZ	8	MoMLV + HPRT (1,7 kb)	+*	+*	+
pH2LacZ	6	H2 Kb (2 kb)	0,02	0,02	+
pMMuLV IL2 LacZ	5	MoMLV + IL-2 (2 kb)	0,01	0,006	-
pMMuLV RAch nlsLacZ	8	MoMLV + RAch (3,3 kb)	nd	nd	-

Table 1 : Activity of various promotors in LacZ constructs introduced in EC cells and 2-cell stage embryo.
The structure of the vector can be found in the indicated references :
1 : Bonnerot *et al* 1987 ; 2 : Hall *et al* 1983 ; 3 : Kalderon *et al* 1984 ; 4 : Sanes et al., 1986 ; 5 : B. Malissen, personal communication ; 6 : E. Wagner, personal communication. 7 : Bonnerot and Nicolas unpublished. See Sassone-Corsi *et al* 1983 for the promotor. 8 : Bonnerot and Nicolas, unpublished.
The expression in EC cells (PCC3 and PCC4, two multipotential cell lines) is calculated as the fraction of the number of positive cells by the number of positive 3T3 or WOP cells obtained from the same calcium phosphate precipitate. +* means positive by infection. yes means positive but not normalized on 3T3. - means undetectable.
The expression in 2-cell stage embryo is obtained by enumerating (data not shown) the number of positive embryos derived after microinjection of supercoiled DNA (with 300 to 3000 copies) into fertilised eggs; + means 10 to 40 % positive embryos, - means less than 1 % (Briand, Grimber, Bonnerot, Nicolas, unpublished).
Abbreviations used for the promotors. SV40e : early promotor of SV40 ; MoMLV : promotor of Moloney murine leukaemia virus ; RSV : promotor of Rous sarcoma virus ; ß-act : promotor of the rat ß-actin gene ; HPRT : promotor of the hypoxanthine-phospho-ribosyl transferase gene ; H-2 K^b : promotor of the class I histocompatibility H-2 K^b gene ; IL-2 : promotor of the interleukin 2 gene ; RAch : promotor of the α–subunit of the acetylcholine receptor gene.

suggesting that they correspond to a late state of multipotential cells. Thus, although information obtained with these cells are very precious, the analysis of promotors must necessarily also be carried out on the embryo.

The test we have utilized uses LacZ as a reporter for promotor functions. Constructs are microinjected into fertilized eggs and analyzed later in development. At the 1-cell stage, no construct is expressed suggesting a general blockage -may be at the transcriptional level- of gene expression. In contrast, at the 2-cell stage, a number of constructs are expressed (Table 1) but a specificity already exists since promotors of specialized genes (the receptor of acetyl-choline, R Ach , or the promotor of the Interleukin II , IL 2) are inactive. When the pattern of expression of a number of promotors is compared in the 2-cell stage and EC, it is evident that these two types of cells behave differently. Two noticeable differences are the expression of the H 2 K^b promotor (pH2LacZ) and the high activity of the 21bp repeat upstream promotor SV 40 element (pA56nlsLacZ). This suggests that the set of *trans*-modulators at these two states of differentiation are distinct (Bonnerot, Briand and Nicolas, unpublished).

These results indicate how careful one must be in the choice of a promotor to drive the expression of a gene so that one obtains either targeted or ubiquituous expression. Further experiments carried out during the pre-implantation period are required before deciding which construct would better fit the desirable properties.

INDUCED GENETIC MOSAICISM

Genetic mosaicism is one of the most powerful techniques used to perform clonal analysis (Garcia Bellido and Merriam,1969). In mammals, this approach however is extremely limited because of the rare nature of spontaneous mutation (Rossant,1987). As we discussed previously, dRRV have properties that make them very attractive as a way to introduce stable genetic modifications *in vivo* and thus to induce genetic mosaicism. There would be three major applications of this approach , firstly as a non invasing way of labelling cells it can be used for clonal and cell lineage analysis. Secondly it could provide a rapid method to obtain answers on the expression of engineered genes without going through the long process of constructing transgenic mice.Thirdly, it may be adapted to the study of genes whose expression is lethal early during embryogenesis. In this paper, we will however concentrate our discussion on only the first of these aspects.

The following conditions are required to provide retroviral induced mosaic mice for clonal and lineage analysis : (1) that the cells whose lineage is to be studied are infectable. The limitations due to this factor are difficult to estimate due to the lack of precise information on the expression of the viral receptor and of the endogeneous gp70 (which may cause interference) during development, (2) that the stock of dRRV is not contaminated by wild type particles. This condition requires the dRRV to be produced in a helper free packaging cell line and to carefully estimate the degree of leakage of the Ψ^- M MuLV RNA in particles. This can be done by using an indicator cell line such as FG 10 (F+L- test, Fischinger et al., 1974) following a step of amplification in 3T3. 3T3-LacZ indicator cell lines in which a complementary LacZ dRRV has been transfected, can also be used for this purpose (by measuring the rescued LacZ dRRV, Rocancourt and Nicolas, unpublished) (3) that the infected cell or some of its progenitors do not complement the defective virus. This imposes some restrictions on the strain of mice to be used since a number of strains develop viremie with age (reviewed in Risser and Horowitz,1983). Careful control is also necessary as it remains possible that even in non-viremic animals, *trans* complementation (without production of wild type virus) occurs due to the expression of defective copies of MMuLV.

So far very few studies have been carried out during the pre-implantation stages of development. It has however been established that the pluripotent cells of embryos can be infected by dRRV, since several transgenic strains of mice that transmit the integrated provirus to their offsprings have been obtained (Nicolas et al.,1985 ; Jähner et al.,1985 ; VanderPutten et al.,1985 ; Huszar et al.,1985 ; Soriano et al., 1987). The degree of infectivity of cells at each stage of development must now be analysed using dRRV LacZ.

The fact that dRRV LacZ can be used in post-implantation embryos animals for clonal analysis has been definitely proved by analysing embryos infected *in utero* by M-MuLVSVLacZ. Stage 10 - 11 (E7) embryos can also be infected with a workable efficiency. At this stage the head process, the neural groove, the notochord and the primitive streak are just begining to develop. Thus direct precursors of the majority of cell types are presumably still not developed and the body plan is just starting to be excecuted. In a very preliminary set of experiments (Sanes et al., 1986) we looked at the presence of infected ß-galactosidase positive cells in two extra embryonic tissues (the amnion and the yolk sac) and one embryonic organ (the skin) 5 to 6 days after infection (E12 or E 13). Labelled clones were detected in all three organs and in each cases their clonal composition was analysed on histological sections. In the skin, dermal clones were obtained, suggesting that mesodermal cells of the proper embryo can be infected. In the yolk sac the clones were of complex structure, composed of mesothelial, fibroblastic and endothelial cells. This clearly indicates that a precursor of all three cell types must exist (probably in the sheet of mesodermal cells present at E7). By analysing the composition of positive clones obtained by later infection at E9 we found that none infected cells gave rise to all three cells types, suggesting that a developmental restriction has dramatically decreased or eliminated multipotential cells present in the yolk sac at E7. Other examples of detection of multipotential precursor cells and of change of cell fate have been reported (Turner and Cepko,1987).

CONCLUDING REMARKS

In the past few years considerable progress has been made in our comprehension of the specificity of gene transcription in differentiated eukaryotic cells. The principles which emerge are fairly simple : a unique combination of *cis*-acting DNA sequences and *trans*-acting factors mediate the regulation of the transcriptional activity of genes. However the specification of the nature of a cell is determined by a series of binary decisions taken during embryogenesis. In mammals the genes involved in these decisions remain to be identified as well as the way cells transfer information to each other during these events. Since cell position, cell interaction and other means of cellular communication must play a key role in these processes, basic questions such as the lineage of cells, the coherence of growth, the variations in the rates of cell multiplication and so on must be examined in detail. Until recently efficient tools to analyse these parameters in the embryos were not available (see however Gardner,1984 ; West, 1984). In this review we have analysed experiments which have been carried out with an aim to investigate whether retroviruses and marker genes could be combined to provide methods adapted to these goals. Several encouraging conclusions can be drawn : (1) as viral vectors, dRRV can be used to introduce genes in vivo : they can infect a variety of somatic cells and can thus be used to create genetic mosaicism either during embryogenesis or in adults ; they can infect the germ cells in the early embryo and can thus be used to produce transgenic animals. The efficiency of both techniques is reasonable although improvements are expected with further sophistication in producting dRRV. (2) The LacZ gene is a suitable *in situ* cell marker at least in short term experiments when expressed into mammalian cells. Problems of background activity in certain cell types can be overcome by using the form of the enzyme targeted towards the nuclear membrane (nls LacZ). It has been used in dRRV, as a genetic marker for the detection of precursors of certain cell types *in vivo* and, tentatively, for the detection of developmental events involved in the restriction of the fate of cells. (3) The LacZ marker has been proved to be particularly efficient as a reporter gene so that the spatial and temporal activity of promotors and the transcriptional activity of the cells can be followed at early stages of development .

There are, however, problems that require further investigation before extending the use of these tools in new directions and in particular before being able to create induced genetic mosaicism, at will, at any stage of development and before having the markers constitutively expressed in transgenic animals.

Firstly, in order to be conclusive, the studies involving the manipulation of dRRV as a marker gene must be carried out in conditions where there is no possibility of reinfecting cells after the first round of replication. Only dRRV viral stocks whose degree of leakiness has been measured and animals (or cells in culture) strictly defective for retroviral complementation must be used. In addition, this problem can be minimized by using dRRV defective for viral driven transcription, since the cells infected with these viruses are unable to produce packagable RNA (Fig.2 C).

Secondly, the ideal vector to introduce genes *in vivo* would have to be neutral for the expression of the inserted gene at all stages of development. Because the LTRs contain complex transcriptional signals in addition to regions involved in others steps of the viral cycle, conventional dRRV cannot be used as such vectors. Again, the vectors which approach nearest to this ideal situation are the dRRVΔEnh in which an internal promotor has been added. It remains to be discovered whether what is left of the virus in the proviral state will be considered as neutral DNA sequences for the insert at all stages of development, a condition required not only for the purpose of cell lineage analysis but also to obtain the targeting of expression by the control of internal tissue-specific promoters.

Thirdly, if the dRRV is used for single cell marking its internal promotor must have the property of ubiquitous expression and it must be expressed at all stages of the cell cycle, as cells in the animal are often stopped in G1 or G0. Viral promotors previously thought to be adapted to this goal have now been found to be too complicated to be able to resolve this problem. As we mentioned previously, several observations about their expression are surprising : for example, although the SV40 early promotor behaves like an ubiquitous promotor in cultured cells and is expressed at the 2-cell stage embryo, its expression in transgenic animals is limited to the cells of the plexus choroid ; the M-MuLV LTR also behave like an ubiquitous promotor in cultured cells but is silent in the 2-cell stage embryo and remain so in transgenic animals. The simplest tool would be the promotors of housekeeping genes.These are now being analysed both in the adult and during embryogenesis.

Acknowledgements : We thank P.Chambon, D.Emilie,T.Friedman, A.Klarsfeld,B.Malissen and E.Wagner for providing plasmids. We would like also to thank P.Briand, G.Grimber, T.Kadesch, C.Pinset, and J.Sanes for sharing unpublished data. We are grateful to E.Legouy and G.Butler for critical reading of the manuscript and to D.Rocancourt and J.C.Bénichou for their help with the photographic work. We wish to thank F.Jacob and P.Berg for encouragement and support. JFN would like to express his gratitude to D.Trano for her patience and support during the preparation of the manuscript.This work was supported by CNRS, INSERM and Pasteur Institut. JFN and CB are from INSERM.

REFERENCES

Bonnerot C, Rocancourt D, Briand P, Grimber G,Nicolas JF (1987). A ß-galactosidase hybrid protein targeted to nuclei as a marker for developmental studies. Proc Nat Acad Sci USA 84:in press.

Coffin J (1985). Genome structure. In Weiss R, Teich N, Varmus H, Coffin J (eds): "RNA Tumor Viruses: Molecular Biology of Tumor Viruses," Cold Spring Harbor Laboratory, p 17.

Dailey L and Basilico C (1985).Sequences in the Polyomavirus DNA Regulatory Region Involved in Viral DNA Replication and Early Gene Expression J of Virology 53:739.

Davidson I, Fromental C, Augereau P, Wildeman A, Zeake M, Chambon P (1986). Cell-type specific protein binding to the enhancer of simian virus 40 in nuclear extracts. Nature 323:544.

Dewey MJ, Mintz B (1978). Direct visualization, by ß-galactosidase histochemistry, of differentiated normal cells derived from malignant teratocarcinoma in allophenic mice. Dev Biol 66:550.

Fischinger PJ, Nomura S, Fuller TN, Dunn KJ (1974). Revertants of mouse cells transformed by murine sarcoma virus. Virology 59:217.

Garcia-Bellido A and Merriam J.R.(1969). Cell lineage in the imaginal disk in *Drosophila* Gynandromorphs.The J of Exp Zool 170:61

Gardner RL, (1984). Mammalian Chimeras-Future Perspectives.Le Douarin N. and McLaren A.(eds):Chimeras in Developmental Biology. Acad Press 431

Gardner RL, Lawrence PA (1985). Single cell marking and cell lineage in animal development. Phil Trans R Soc Lond B 312:1-187.

Gilboa E, Mitra SW, Goff S, Baltimore D (1979). A detailed model of reverse transcription and tests of crucial aspects. Cell 18:93.

Goring DR, Rossant J, Clapoff S, Breitman ML, Tsui LC (1987). In situ detection of ß-galactosidase in lenses of transgenic mice with a -crystallin/lacZ gene. Science 235:456.

Gorman CM, Merlino GT, Willingham MC, Pastan I, Howard BH

(1982). The Rous sarcoma virus long terminal repeat is a strong promoter when introduced into a variety of eukaryotic cells by DNA-mediated transfection. Proc Nat Acad Sci USA 79:6777.

Gorman CM, Rigby PW, Lane DR (1985). Negative regulation of viral enhancers in undifferentiated embryonic stem cells. Cell 42:519.

Hall CV, Jacob PE, Ringold GM, Lee F (1983). Expression and regulation of Escherichia coli lacZ gene fusions in mammalian cells. J Mol Appl Genet 2:101.

Herbomel P, Bourachot B and Yaniv M(1984). Two distinct Enhancers with different Cell specificities coexist in the regulatory region of Polyoma. Cell 39:653.

Huszar D, Balling R, Kothary R, Magli MC, Hozumi N, Rossant J, Bernstein A (1985). Insertion of a bacterial gene into the mouse germ line using an infectious retrovirus vector. Proc Nat Acad Sci USA 82:8587.

Jähner D and Jaenisch R (1980). Integration of Moloney leukemia virus into the germ line of mice: correlation between site of integration and virus activation. Nature 287:456.

Jähner D, Haase K, Mulligan R, Jaenisch R (1985). Insertion of the bacterial gpt gene into the germ line of mice by retroviral infection. Proc Nat Acad Sci USA 82:6927.

Jakob H, Nicolas JF (1987). Mouse teratocarcinoma cells. Methods in Enzymol 151:66.

Kalderon D, Roberts BL, Richardson WD, Smith AE (1984). A short amino acid sequence able to specify nuclear location. Cell 39:499.

Kauffman S (1975). Control circuits for determination and transdetermination : interpreting positional information in a binary epigenetic code. In Elsevier (ed) :"Cell patterning," Ciba Foundation Symposium 29 (new series) Amsterdam, p 201.

Kelly F, Condamine H (1982). Tumor viruses and early mouse embryos. Biochim Biophys Acta 651:105.

Linney E, Davis B, Overhauser J, Chao E and Fan H.(1984). Non-function of a Moloney murine leukaemia virus regulatory sequence in F9 embryonal carcinoma cells . Nature 308:470

Lo CW (1986). Localization of low abundance DNA sequences in tissue sections by in situ hybridization. J Cell Sci 81:143.

Maniatis T, Goodbourn S, Fischer JA (1987). Regulation of inducible and tissue-specific gene expression.Science 236:1237.

Mann R, Baltimore D (1985). Varying the position of a retrovirus packaging sequence results in the encapsidation of both unspliced and spliced RNAs. J Virol 54:401.

Mann R, Mulligan RC, Baltimore D (1983). Construction of a retrovirus packaging mutant and its use to produce helper-free defective retrovirus. Cell 33:153.

McKnight S, Tjian R (1986). Transcriptional selectivity of viral genes

in mammalian cells. Cell 48:795.
Miller AD, Buttimore C (1986). Redesign of retrovirus packaging cell lines to avoid recombination leading to helper virus production. Mol Cell Biol 6:2895.
Mulligan RC, Berg P (1980). Expression of a bacterial gene in mammalian cells. Science 209:1422.
Nicolas JF, Berg P (1983). Regulation of expression of genes transduced into embryonal carcinoma cells. In Silver LM, Martin GR, Strickland S (eds) :"Teratocarcinoma Stem Cells," Cold Spring
Nicolas JF, Rubenstein JLR (1987). Retroviral vectors. In Rodriguez RL, Denhardt DT (eds) :"Vectors: A Survey of Molecular Cloning Vectors and Their Uses," Butterworth Publ.Co, Stoneham Massachusetts, in press.Harbor Conferences on Cell Proliferation 10:469.
Nicolas JF, Jakob H, Jacob F (1981). Teratocarcinoma-derived cell lines and their use in the study of differentiation. Sato G (ed) : "Functionally differentiated cell lines," Alan R Liss Publ, New York, p 185.
Nicolas JF, Avner P, Gaillard J, Guenet JL, Jakob H and Jacob F (1976). Cells lines derived from Teratocarcinomas.Cancer Research 36:4224
Nicolas JF, Rubenstein JLR, Bonnerot C, Jacob F (1985). Introduction of genes into embryonal carcinoma cells and preimplantation embryos by retroviral vectors. Cold Spring Harb Symp Quant Biol 50:713.
Nudel U, Zakut R, Shani M, Neuman S, Levy Z and Yaffe D (1983).The nucleotide sequence of the rat cytoplasmic ß-actin gene. Nucleic Acid Research.11:1759.
Palmiter RD and Brinster RL (1986). Germ-line transformation of mice. Ann Rev Genet 20:465.
Price J, Turner D, Cepko C (1987). Lineage analysis in the vertebrate nervous system by retrovirus-mediated gene transfer. Proc Natl Acad Sci USA 84:156.
Risser R, Horowitz J (1983). Endogenous mouse leukemia viruses. Ann Rev Genet 17:85.
Rossant J (1987). Cell lineage analysis in mammalian embryogenesis. Curr Top Dev Biol, in press.
Rubenstein JLR, Nicolas JF, Jacob F (1984). Construction of a retrovirus capable of transducing and expressing genes in multipotential embryonic cells. Proc Natl Acad Sci USA 81:7137.
Sanes JR, Rubenstein JLR, Nicolas JF (1986). Use of a recombinant retrovirus to study post-implantation cell lineage in mouse embryos. The EMBO Journal 5:3133.
Sassone-Corsi P, Duboule D, Chambon P (1985). Viral enhancer activity in teratocarcinoma cells. Cold Spring Harb Symp Quant

Biol 50:747.
Schirm S, Jiricny J, Schaffner W (1987). The SV40 enhancer can be dissected into multiple segments, each with a different cell type specificity. Genes & Development 1:65.
Sen R, Baltimore D (1986). Multiple nuclear factors interact with the immunoglobulin enhancer sequences. Cell 46:705.
Soriano P, Cone RD, Mulligan RC, Jaenisch R (1987). Tissue specific and ectopic expression of genes introduced into transgenic mice by retroviruses. Science, in press.
Taketo M, Gilboa E, Sherman MI (1985). Isolation of embryonal carcinoma cell lines that express integrated recombinant genes flanked by the Moloney murine leukemia virus long terminal repeat. Proc Natl Acad Sci USA 82:2422.
Turner DL, Cepko CL (1987). A common progenitor for neurons and glia persists in rat retina late in development. Nature 328:131.
Van Dyke T, Finlay C, Levine AJ (1985). A comparison of several lines of transgenic mice containing the SV40 early genes. Cold Spring Harb Symp Quant Biol 50:671.
Van der Putten H, Botteri FM, Miller AD, Rosenfeld MG, Fan H, Evans RM, Verma IM (1985). Efficient insertion of genes into the mouse germ line via retroviral vectors. Proc Natl Acad Sci USA 82:6148.
Wagner EF, Vanek M, Vennström B (1985). Transfer of genes into embryonal carcinoma cells by retrovirus infection : efficient expression from an internal promoter. The EMBO Journal 4:663.
West JD,(1984).Cell markers. LeDouarin N and McLaren A. (eds) : Chimeras in Developmental Biology. Acad Press 39.
Yu SF, von Rüden T, Kantoff PW, Garber C, Seiberg M, Rüther U, Anderson WF, Wagner EF, Gilboa E (1986). Self-inactivating retroviral vectors designed for transfer of whole genes into mammalian cells. Proc Natl Acad Sci USA 83:3194.

Cellular Factors in Development and Differentiation:
Embryos, Teratocarcinomas, and Differentiated Tissues, pages 147–161

ANALYSIS OF A LETHAL MUTATION (t^{w18}) IN THE MOUSE *t*-COMPLEX

Gail R. Martin, Alexandra Joyner, Lee M. Silver, Hans Lehrach and Maja Bucan

Department of Anatomy, (G.R.M., A.J.), University of California, San Francisco, CA 94143; Department of Molecular Biology (L.M.S.), Princeton University, Princeton, NJ 08540; Imperial Cancer Research Fund (H.L., M.B.), Lincoln's Inn Fields, London WC2A 3PX, UK

INTRODUCTION

A significant proportion of house mice (Mus musculus and Mus domesticus) trapped in the wild carry an abnormal form of the proximal portion of chromosome 17 (reviewed by Silver, 1985). The region of the genome that is altered in these animals is now termed the "*t*-complex." It extends over a distance of approximately 15cM and includes the major histocompatibility complex (MHC or H-2 complex). The variant form of the *t*-complex that is found in wild mice is known as a *t*-haplotype.

Properties of *t*-haplotypes.

Almost all *t*-haplotypes isolated from natural populations have four unusual properties in common. The variant chromosome segment:

1) is unable to recombine normally with its wild-type counterpart;

2) contains a group of mutant genes, distributed throughout the *t*-complex that interact to cause transmission ratio distortion (TRD) from heterozygous (+/*t*) males, such that the *t*-haplotype is transmitted to virtually all offspring;

3) contains a group of genes, presumably the same as those that cause TRD, that interact to cause sterility in males carrying two *t*-haplotypes (*t/t*);
4) is usually associated with a recessive embryonic lethal mutation.

These recessive lethal mutations fall into a number of complementation groups, and animals heterozygous for two different lethal *t*-mutations can be viable. At present 16 different complementation groups are known (Klein, et al., 1984). Although few of these have been studied in any detail, it appears that each member of a particular complementation group causes the same distinctive syndrome of disturbances leading to embryonic death, and that members of the different groups interrupt development at different times (Bennett, 1975; Sherman and Wudl, 1977). It is by virtue of these lethal mutations that *t*-haplotypes traditionally have been distinguished from one another and classified.

Until recently, it was not clear whether the various properties associated with *t*-haplotypes had any relationship to one another. However, with an increased understanding of the genetic and molecular structure of *t*-haplotype-bearing chromosomes and the genetic basis for *t*-haploptype effects on sperm function, it became possible to formulate a unifying hypothesis that accounts for all of the properties characteristic of *t*-haploytpes. First, an explanation for the almost complete suppression of recombination between *t*-haplotype and wild-type chromosomal segments has been provided by the demonstration that two regions of the *t*-complex, one in the proximal and one in the distal portion, are inverted in *t*-haplotypes (Artzt et al., 1982; Pla and Condamine, 1984; Herrmann et al., 1986; see Figure 1). Second, it is now apparent that the suppression of recombination due to these inversions is responsible for holding all of the TRD genes together from one generation to the next, and that the TRD that they cause is responsible for the spread and maintenance of *t*-haplotypes at relatively high frequencies in natural populations. Third, sterility may be an unavoidable consequence of homozygosity for the complete set of TRD genes (Lyon, 1986). Finally, recessive lethal mutations prevent the birth of sterile males that would otherwise compete with fertile +/*t* males for food, breeding territories and mates.

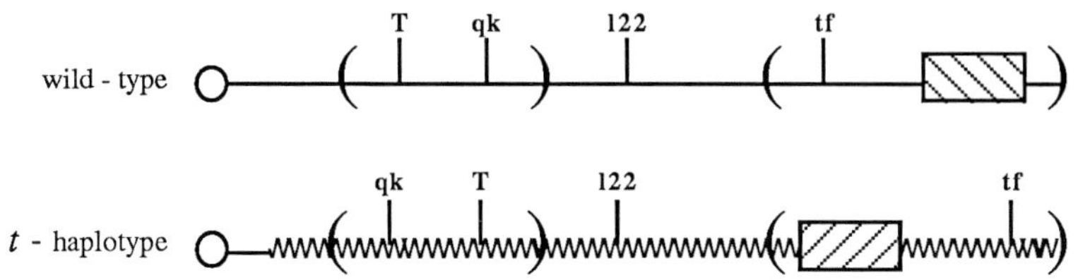

Figure 1. Structure of the *t*-complex. The circle represents the centromere of chromosome 17. Wild-type and *t*-haplotype chromatin are indicated by straight and zig-zag lines, respectively. The parentheses () indicate the positions of regions that are inverted. The positions of several phenotypic markers are shown: T, a locus affecting tail-length, quaking (qk), and tufted (tf). Also shown are the D17Leh122 locus (122), detected with the Tu122 DNA probe, and the H-2 complex (cross-hatched box). The distance between markers is not drawn to scale

According to this view, the lethal alleles are due to random mutations in unrelated genes that occurred during evolution and were retained because they provided a selective advantage to *t*-haplotypes that carry them over those that lack them. An alternative view has been suggested by Bennett (1975) and her colleagues (Shin, et al., 1983), who propose that the genes represented by the lethal alleles found in *t*-haplotypes are all related to one another and have a common function, perhaps encoding cell surface antigens or other types of molecules that might mediate cell-cell interactions in embryonic development.

Whatever their relationship to one another, the existence of the *t*-complex lethal mutations has served to identify genes whose functions are required during mammalian development, several of which appear to play a role during the first third of gestation. Relatively few mutations other than those in the *t*-complex are known to affect this period, and therefore embryos homozygous for certain lethal *t*-haplotypes provide valuable experimental material for studies of genes that affect early embryonic development.

Approaches to Studying Lethal Mutations in the *t*-complex.

Most attempts to go beyond strictly morphological analysis of lethal *t*-mutation effects have involved studies of the behavior of mutant embryos grafted to extra-uterine sites in vivo or in vitro. Interpretation of the data from such studies was severely hampered by the lack of markers that could unequivocally distinguish mutant homozygotes (*t/t*) from their heterozygous (+/*t*) and wild-type (+/+) littermates. Thus, investigators were forced to rely upon statistical analysis of their data to determine whether a particular abnormal phenotype attributed to *t/t* embryos occured at the expected frequency. Based on such analyses, two general hypotheses have been proposed for the mode of action of the different early acting *t*-haplotype lethal mutations. One is that each mutation is an "organizational lethal," which acts to produce abnormal cell-cell interactions, and that embryonic death is a secondary consequence of tissue disorganization and death of individual cell types within the embryo (Bennett, 1975). An alternative view is that at least some of the mutations are "generalized cell lethals," which cause embryonic death by perturbing processes such as intermediary metabolism that are essential for the survival of all cells in the embryo (Wudl, et al., 1977).

One approach to distinguishing between these alternatives is based on the finding that it is possible to establish a pluripotent embryonic stem cell (ESC) line from the inner cell mass (ICM) of a mouse blastocyst (Martin, 1981; Evans and Kaufman, 1981). Under appropriate conditions such cells can be induced to differentiate in a manner that mimics the behavior in vitro of the peri-implantation embryo in utero, and when injected subcutaneously into mice, these cells form tumors (teratocarcinomas) that contain a variety of differentiated cell types. When ESC lines can be obtained from embryos homozygous for a given lethal *t*-haplotype, this provides strong evidence that the lethal mutation is not a "generalized cell lethal." Since a number of markers for *t*-haplotypes have been identified, including a set of cloned DNA sequences that recognize restriction fragment length polymorphisms (RFLPs) between *t*-haplotype and wild-type DNA (Roehme, et al., 1984; Fox, et al.,1985), it is therefore now possible to determine unequivocally the genotype of each ESC line, and thus to use the technique of ES cell isolation to determine the viability of mutant homozygous embryonic cells.

We have used this approach to obtain evidence that it is possible to establish pluripotent ESC lines homozygous for the t^{w5} lethal mutation (Magnuson, et al., 1982; Martin, et al., 1987), thus contradicting the suggestion that t^{w5} is a "generalized cell lethal" (Wudl and Sherman, 1976). We have further obtained evidence (Martin, et al., 1987) consistent with the suggestion that the normal function of the t^{0} lethal gene is required for the survival of the cells in the mouse blastocyst (Wudl, et al., 1977).

This paper describes the results of our recent study (Martin, et al., 1987) aimed at applying this type of analysis to an early-acting lethal mutation, t^{w18}, which is a member of the t^{9} complementation group. We also describe here the results of subsequent studies, employing ESC lines derived from embryos homozygous for the t^{w18}-haplotype, aimed at elucidating the molecular structure of members of the t^{9} complementation group (Bucan, et al., 1987).

RESULTS AND DISCUSSION

Establishment of t^{w18}/t^{w18} ESC Lines.

The approach we took was to derive ESC lines from embryos obtained by <u>inter-se</u> mating of $+/t^{w18}$

animals, and to determine the genotype of each cell line established by analysis of at least one of four genetic markers. ICMs from a total of 629 embryos were cultured, and of these, 334 (53%) formed primary stem cell colonies. Of these, 29/312 (9.3%) gave rise to ESC lines: 3 were found to be homozygous for the t^{w18} haplotype, 16 were found to be $+/t^{w18}$ heterozygotes, and 10 were found to be wild-type. Figure 2 illustrates the type of molecular analysis of the genotypes of the three t^{w18}/t^{w18} ESC lines that was performed. The results of this analysis show that each of the three cell lines is homozygous for the *t*-specific allele of the D17Leh122 locus, represented by a 1.9 kb TaqI DNA fragment.

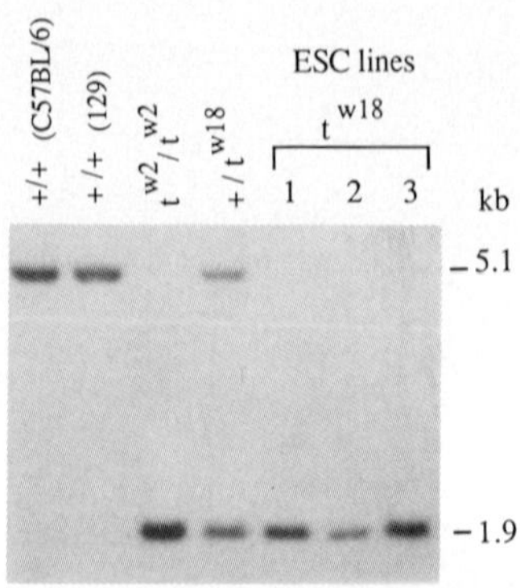

Figure 2. Southern blot hybridization analysis of ESC lines. TaqI restriction fragments of genomic DNA isolated from the spleens of adult animals and from three ESC lines were separated by electrophoresis, transferred to nylon membranes and hybridized with the Tul22 probe (Fox,et al., 1985). Spleens were obtained from wild-type animals (strains 129 and C57BL/6), from a male homozygous for the t^{w2} haplotype (a semi-lethal complete *t*-haplotype), and a female heterozygous for the t^{w18} lethal haplotype. The three ESC lines were derived from embryos obtained by *inter-se* mating of $+/t^{w18}$ animals.

The expected number of t^{w18} homozygous, heterozygous and wild-type cell lines can be calculated by taking into account the distorted ratio of transmission of the *t*-haplotype in males. Like all other members of the t^9 complementation group, t^{w18} is known to be a "partial" *t*-haplotype, produced by a rare recombination event in the central portion of the *t*-complex. The proximal portion of *t*-haplotype chromatin, including the D17Leh122 locus, has been retained in the t^{w18}-haplotype, but the distal portion, which includes one of the genetic TRD factors (Tcd-2), has been exchanged for wild-type chromatin. Consequently, t^{w18} is not transmitted at the exceptionally high ratios usually observed for complete *t*-haplotypes such as t^{w5} (Lyon, 1984). In our study the average transmission ratio of the t^{w18} haplotype from six parental males was found to range from 0.53 to 0.66, with an average of 0.59. If one assumes an equal probability of obtaining ESC lines from *t/t*, +/*t* and +/+ embryos, one would expect that 8.55 of the 29 ESC lines obtained would have been homozygous for the t^{w18} lethal haplotype. However, only 1/3 of the expected number of t^{w18} homozygotes was obtained. These data demonstrate that embryonic cells homozygous for the t^{w18} lethal haplotype are capable of giving rise to ESC lines, but suggest that the growth of t^{w18}/t^{w18} ICM cells in vitro may be impaired.

The differentiative capacity of each of the t^{w18}/t^{w18} ESC lines was analyzed in vitro and/or in vivo. When cultured under appropriate conditions, these cell lines formed free-floating embryo-like structures (embryoid bodies) similar to those formed by pluripotent, wild-type cells (Martin, 1981). The multi-layered cultures that developed following attachment of such embryoid bodies to a substratum contained a variety of morphologically distinct cell types. Although most differentiated cell types were difficult to identify in these complex cultures, neuronal cells with long extended processes were discernable and muscle could be readily detected when it occurred in patches of rhythmically contracting cells. Specific differentiated cell types, including keratinizing and pigmented epithelium, cartilage, muscle and adipose were more readily identified in histological sections of the tumors that formed following subcutaneous injection into athymic mice (Figure 3).

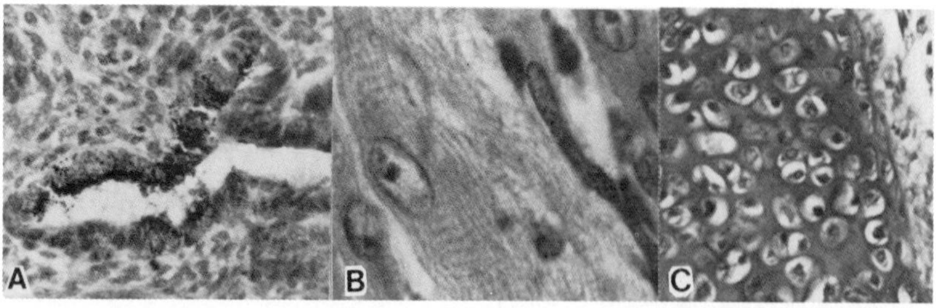

Figure 3. Histological analysis of tumors formed by t^{w18}/t^{w18} ESC lines in athymic mice. (A) a region containing an epithelial structure with black pigment granules; (B) a region containing multinucleate, striated muscle cells; (C) a region containing cartilage.

These data provide some insight into the defects engendered by homozygosity for the t^{w18} lethal haplotype. Morphological studies of embryos obtained by mating mice heterozygous for the t^{w18} haplotype, or other independently-derived alleles in the same complementation group, have indicated that mutant homozygotes are characterized by an accumulation of cells in the region of the primitive streak at 7.5 days of gestation, and by the subsequent lack of normal mesodermal structures (Bennett and Dunn, 1960; Moser and Gluecksohn-Waelsch, 1967). Based on ultrastructural studies of cells in the primitive streak region of presumed mutant homozygotes, it was suggested that homozygosity for the t^{w18} lethal gene causes a block to mesodermal differentiation (Speigelman and Bennett, 1974). Consistent with this was the finding by Artzt and Bennett (1972) that 7.5 and 8.5 day presumed t^{w18}/t^{w18} embryos transplanted to the testis formed growths lacking mesodermal derivatives. An alterative view comes from a study of mitosis in presumed mutant homozygotes at 6.25 - 7.5 days of gestation (Snow and Bennett, 1978). Their data suggest that the precursors of mesoderm in the epiblast of mutant embryos undergo mitotic arrest, whereas other cells remain unaffected. The results of the study by Kelly, et al. (1979) are consistent with the latter hypothesis,that mitosis is impaired in mutant homozygotes.

Our data show that t^{w18}/t^{w18} ES cell lines can form typical teratocarcinomas, similar to those formed by wild-type ESC lines, with no obvious reduction in the proportion of mesodermal derivatives. This indicates that the capacity to form mesodermal derivatives is not impaired in t^{w18}-homozygous ES cells. We therefore conclude that the normal function of the gene(s) represented by the t^{w18}-lethal mutation is not involved in the specification of the mesodermal lineage. Instead, we hypothesize that its function is required for the survival of mesodermal precursors in the embryo. Moreover, the lethal effect of the t^{w18} haplotype on such cells is manifested only in the highly structured environment of the early post-implantation embryo, and not when differentiation is disorganized in vitro or in vivo.

Our finding that t^{w18}/t^{w18} ESC lines were obtained at a frequency approximately three-fold lower than expected might be due to the proposed defect in mitosis. If so, it is not clear why 30% of t^{w18}/t^{w18} embryos can escape the effects on mitosis. In view of our finding that the rare recombination event that created the t^{w18} haplotype resulted in apparently large deletions and duplications of DNA sequences (see below), it is possible that the low frequencies at which we obtained t^{w18}/t^{w18} ESC lines may be the consequence of deletion and/or duplication of *t*-complex DNA sequences in the t^{w18}-haplotype that are unrelated to the embryonic lethal mutation hypothesized to cause mitotic arrest in mesodermal precursors.

Molecular Analysis of the t^9 Complementation Group

Among the lethal *t*-mutations, members of the t^9 complementation group are exceptional in that none have been found in wild populations. Instead, they have all arisen in laboratory stocks by recombination between *t*-haplotype and wild-type DNA, with only six alleles identified in over 50 years of laboratory breeding: t^9 and t^4 (Dunn and Gluecksohn-Waelsch, 1953; Moser and Gluecksohn-Waelsch, 1967), t^{w18} (Bennett and Dunn, 1960), t^{w30} (Dunn et al., 1962), t^{w52} (Bennett et al., 1976) and t^{ks1} (Bucan, et al., 1987). In light of the evidence that recombination between *t*-haplotype and wild-type DNA in the proximal portion of the *t*-complex can result in deletion and/or duplication of genetic material (Herrmann, et al., 1987; Sarvetnick, et al., 1986), we sought to determine whether the members of the t^9 complementation group contain similar genomic alterations, which might be responsible for the lethal phenotype common to these *t*-haplotypes.

This investigation relied on the use of random DNA clones originally isolated from the proximal portion of mouse chromosome 17 by microdissection and microcloning (Roehme et al., 1984). Twenty-five of these cloned DNA probes were selected on the basis of the finding that they represent unique or low copy number sequences that map to chromosome 17. To determine whether any of these sequences are deleted in the t^{w18} haplotype, we took advantage of the availability of the t^{w18}/t^{w18} ESC lines described above. Genomic DNA was isolated from these ES cells, as well as from ES cells homozygous for a complete *t*-haplotype, t^{w5}, and from wild-type ES cells (Martin, et al., 1987), digested with restriction endonucleases and analyzed by the Southern blot hybridization technique.

Clone Tu443, which hybridizes to a single MspI fragment of approximately 15 kb in wild-type DNA and of 8kb in t^{w5}/t^{w5} DNA, did not hybridize to any MspI fragment in DNA from t^{w18}/t^{w18} cells (Figure 4A). Control experiments were performed to demonstrate that this lack of hybridization was not due to the absence of DNA fragments in the lane. Similar results were obtained with three other probes, providing clear evidence that the D17Leh443, 94, 54 and 180 loci identified by the Tu443, Tu94, p54M and Tu180 probes, respectively, are deleted in the t^{w18} haplotype. These four loci were found to lie in a cluster approximately 1 cM proximal to tf in wild-type. At least two of these loci, D17Leh443 and

54, were also found to be deleted in two other members of the t^9 complementation group, t^4 and t^{ks1}.

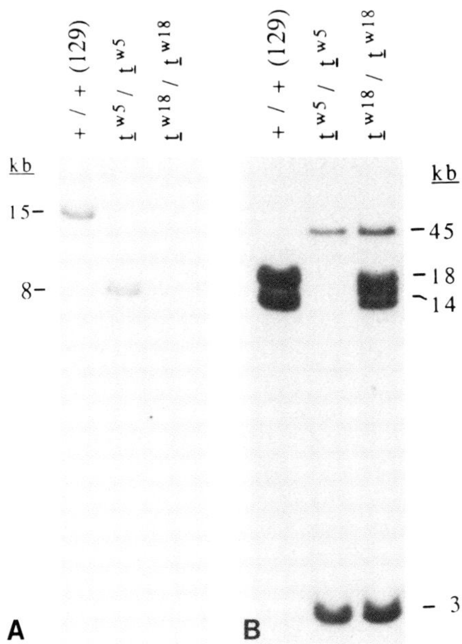

Figure 4. Deletion and duplication of sequences in the t^{w18} haplotype. Southern blot hybridization analysis of genomic DNA from cells or animals of the designated genotypes. Each lane contains 10 μg of genomic DNA digested with (A) MspI and was hybridized with the Tu443 probe; (B) TaqI and was hybridized with the Tu89 probe.

To determine whether sequences are duplicated in the t^{w18} haplotype, an analysis was carried out with a subset of the remaining probes, which detect RFLPs between wild-type and *t*-haplotype DNA. The Tu89 probe hybridizes strongly to three MspI fragments of approximately 14-18 kb in wild-type DNA and to two MspI fragments of 45 and 3 kb in DNA from t^{w5}/t^{w5} cells. When the Tu89 probe was hybridized to MspI digests of t^{w18}/t^{w18} DNA, the hybridization pattern observed indicated that the t^{w18} haplotype contains the three wild-type-specific and the two *t*-specific Tu89 fragments (Figure 4B). It thus appears that the D17Leh89 sequences detected by the Tu89 probe are duplicated in the t^{w18} haplotype. Similarly evidence was obtained that the D17Leh467 and D17Leh525 sequences are also duplicated in the t^{w18} haplotype. Further analysis revealed that these loci lie in a cluster that maps distal to the Qa-2 region of the H-2 complex in wild-type, or proximal to this region in *t*-haplotypes (Bucan, et al., 1987, see Figure 5). All three loci are also duplicated in the t^4 and t^{ks1} haplotypes.

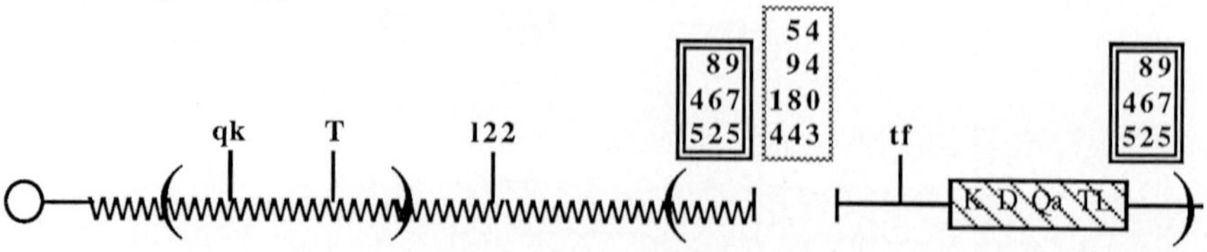

Figure 5. Genetic map of members of the t^9 complementation group. The circle indicates the centromere of chromosome 17. The straight and zig-zag lines represent wild-type- and *t*-haplotype-derived chromatin, respectively. The parentheses () indicate the positions of regions that are inverted in *t*-haplotypes, relative to wild-type. The relative order of the markers is shown. Loci present in single copy: qk, T, D17Leh122, tf, H-2 complex (cross-hatched box). Loci deleted: D17Leh54, 94, 180, and 443. Loci duplicated: D17Leh89, 467 and 525.

From our data it is not possible to obtain a minimal estimate for the size of the deletion in the members of the t^9 complementation group, since no recombination between any of the deleted markers was detected in the products of a high resolution genetic cross (Herrmann, Bucan and Lehrach, unpublished). However, the fact that the sequences represented by 4 probes out of 25 tested were found to be contained in the deletion might be an indication of a fairly large size. Similarly, the duplicated region, represented by 3 other probes may be relatively large.

At present it is uncertain whether the deletion or the duplication is responsible for the recessive lethal phenotype characteristic of members of the t^9 complementation group. If the latter is the case, however, then one would have to argue that the detrimental effects of the aneuploidy are manifested only in animals tetrasomic for the region in question, since $+/t^{w18}$, $+/t^4$ and $+/t^{ks1}$ heterozygotes, which are trisomic for that region, are completely normal. Therefore, it seems most likely that the cause of the embryonic lethality in t^{w18} homozygotes is nullisomy for a vital gene. Although further analysis of the molecular nature of the lethal mutation will be difficult, especially considering the potentially large size of the structurally altered regions, the data reported here provide a starting point for the search for the lethal gene.

REFERENCES

Artzt K, Bennett D (1972). A genetically caused ectodermal tumor in the mouse. J Nat Cancer Inst 48:141-158.

Artzt K, Shin H-S, Bennett D (1982). Gene mapping within the *t/t* complex of the mouse. II. Anomalous position of the H-2 complex in *t* haplotypes. Cell 28:471-476.

Bennett D (1975). The T-locus of the mouse. Cell 6:441-454.

Bennett D, Dunn LC (1960). A lethal mutant (t^{w18}) in the house mouse showing partial duplications. J Exp Zool 143:203-219.

Bennett D, Dunn LC, Artzt K (1976). Genetic changes in mutations at the T/t locus in the mouse. Genetics 83:361-372.

Bucan M, Herrmann B, Frischauf A-M, Bautsch V, Silver LM, Bode V, Martin GR, Lehrach H (1987). Deletion and duplication of DNA sequences is associated with the embryonic lethal phenotype of the t^9 complementation group of the mouse *t* complex. Genes and Devel 1:376-385.

Dunn LC, Gluecksohn-Waelsch S (1953). Genetic analysis of seven newly discovered mutant alleles at locus T in the house mouse. Genetics 38:261-271.

Dunn LC, Bennett D, Beasley AB (1962). Mutation and recombination in the vicinity of a complex gene. Genetics 47:285-303.

Evans MJ, Kaufman MH (1981). Establishment in culture of pluripotential cells from mouse embryos. Nature 292:154-156.

Fox HS, Martin GR, Lyon MF, Herrmann B, Frischauf A-M, Lehrach H, Silver LM (1985). Molecular probes define different regions of the mouse t complex. Cell 40:63-69.

Herrmann B, Bucan M, Mains PE, Frischauf A-M, Silver LM, Lehrach H (1986). Genetic analysis of the proximal portion of the mouse *t*-complex: evidence for a second inversion within *t* haplotypes. Cell 44:469-476.

Herrmann BG, Barlow DP, Lehrach H (1987). A large inverted duplication allows homologous recombination between chromosomes heterozygous for the proximal *t* complex inversion. Cell 48:813-825.

Kelly F, Guenet JL, Condamine H (1979). Karyologically identified homozygous t^{w18} embryos: extrauterine growth properties and transformation by SV40. Cell 16:919-927.

Klein J, Sipos P, Figueroa F (1984). Polymorphism of t complex genes in European wild mice. Genet Res 44:39-46.

Lyon MF (1984). Transmission ratio distortion in mouse *t*-haplotypes is due to multiple distorter genes acting on a responder locus. Cell 37:621-628.

Lyon MF (1986). Male sterility of the mouse *t*-complex is due to homozygosity of the distorter genes. Cell 44:357-363.

Magnuson T, Epstein CJ, Silver LM, Martin GR (1982). Pluripotent embryonic stem cell lines can be derived from t^{w5}/t^{w5} blastocysts. Nature 298:750-753.

Martin GR (1981). Establishment of pluripotent cell lines from embryos cultured in medium conditioned by teratocarcinoma stem cells. Proc Natl Acad Sci 78:7634-7638.

Martin GR, Fox HS, Silver LM, Joyner AL (1987). Establishment of embryonic stem cell lines from preimplantation mouse embryos homozygous for lethal mutations in the *t*-complex. Devel Biol 121: 20-28.

Moser GC and Gluecksohn-Waelsch S (1967). Developmental genetics of a recessive allele at the complex T-locus in the mouse. Devel Biol 16:564-576.

Pla M, Condamine H (1984). Recombination between two mouse t haplotypes (t^{w12}tf and t^{Lub-1}): Mapping of the H-2 complex relative to centromere and tufted (tf) locus. Immunogenet 20:277-285.

Roehme D, Fox H, Herrmann B, Frischauf A-M, Edstrom JE, Mains P, Silver LM, Lehrach H (1984). Molecular clones of the mouse *t* complex derived from microdissected metaphase chromosomes. Cell 36:969-976.

Sarvetnick N, Fox HS, Mann E, Mains PE, Elliott RW, Silver LM (1986). Nonhomologous pairing in mice heterozygous for a *t*-haplotype can produce recombinant chromosomes with duplications and deletions. Genetics 113:723-743.

Sherman M, Wudl LR (1977). T-complex mutations and their effects. In: Sherman M (ed): "Concepts in Mammalian Embryogenesis," Cambridge, Mass: MIT Press, pp 136-234.

Shin H-S, McCormick P, Artzt K, Bennett D (1983). Cis-trans tests show a functional relationship between non-allelic mutations in the T/t complex. Cell 33:925-929.

Silver LM (1985). Mouse t haplotypes. Ann Rev Genet 19:179-208.

Snow MHL, Bennett D (1978). Gastrulation in the mouse: assessment of cell populations in the epiblast of t^{w18}/t^{w18} embryos. J Embryol exp Morphol 47:39-52.

Spiegelman M, Bennett D (1974). Fine structural study of cell migration in the early mesoderm of normal and mutant mouse embryos. J Embryol exp Morphol 32:723-738.

Wudl LR, Sherman MI (1976). In vitro studies of mouse embryos bearing mutations at the T locus: t^{w5} and t^{12}. Cell 9:523-531.

Wudl LR, Sherman MI, Hillman N. (1977). Nature of lethality of t mutations in embryos. Nature 270: 137-140.

Cellular Factors in Development and Differentiation:
Embryos, Teratocarcinomas, and Differentiated Tissues, pages 163–186

THE EXPRESSION OF VIRAL AND CELLULAR GENES IN PAPILLOMAS OF THE CHOROID PLEXUS INDUCED IN TRANSGENIC MICE

J. Marks, J. Lin, D. Miller, G. Lozano, J. Herbert and A.J. Levine
Molecular Biology, Princeton University, Princeton, NJ 08544 (J.M., J.L., G.L. & A.J.L.), Pathology & Neuropathology, NYU Medical Center, NY, NY 10016 (D.M.), and Neurology, Columbia University, College of Physicians & Surgeons, NY, NY 10032 (J.H.)

SUMMARY

A line of transgenic mice that carry the SV40 gene for the large Tumor antigen express this protein during the first two weeks of life in brain tissue. By 30-40 days after birth, independently derived multiple foci of abnormal cells appear throughout the choroid plexus. After 90 days, higher levels of T antigen and rapid tumor growth are detected and all these animals die in a narrow time span, between 100-120 days. In situ hybridization with tissue sections and Northern blot analysis have been employed to follow the steady state levels of SV40 RNA and the p53 oncogene RNA levels in normal and tumor tissues. The level of SV40 RNA is quite variable between tumor cells in a section. This heterogeneity of T antigen mRNA levels could permit the selection of cells (from the multiple foci) expressing higher levels of T antigen and growing more rapidly. The increased levels of p53 RNA observed in tumor cells could then result from the active growth state of these cells or a more direct transcriptional activation. Two cellular genes, transthyretin and the 5-HT_{1C} serotonin receptor, both of which are preferentially expressed in normal choroid plexus cells, were also examined for RNA production in these tumors of the choroid plexus. Both of these genes produced high levels of RNA in tumor tissue indicating the retention of well differentiated gene expression in these tumor tissues. This reflects, at the level of gene expression, the well differentiated morphology of these papillomas of the choroid plexus. Interestingly, as cell lines have been derived from these tumors, both the choroid plexus specific RNA species (for 5-HT_{1C} receptor) and characteristic morphology were lost and an increase in T antigen levels was observed.

INTRODUCTION

Transgenic mice carrying the early region genes of the tumor virus Simian Virus 40 (SV40) reproducibly develop a neoplasia of the choroid plexus (Brinster, *et al.*, 1984; Van Dyke, *et al.*, 1987). The choroid plexus is composed of the lining of epithelial cells in the ventricles of the brain. This organ filters the plasma proteins of the blood and produces the cerebrospinal fluid (CSF) that baths the central nervous system. This highly vascularized tissue forms the blood-CSF barrier which is part of the blood-brain barrier. The SV40 large T antigen gene, under the control of the viral enhancer-promoter signals, is necessary to induce a papilloma of the choroid plexus observed in transgenic mice (Palmiter, *et al.*, 1985). The enhancer-promoter region provides the tissue restricted expression of the viral tumor antigen (Brinster, *et al.*, 1984; Palmiter, *et al.*, 1985) and a mutant T antigen fails to produce a tumor demonstrating the essential role of this viral protein in tumorigenesis. All animals of the SV11 line of transgenic mice (Van Dyke, *et al.*, 1986; 1987) succumb to the tumor within a narrow age bracket. Over ten generations of backcrosses, of C57BL6/J X SJL mice to C57BL6/J mice, the mean age of 119 mice at death was 104 $\pm$ 12 days. From this line, 28 mice were sacrificed at ages varying from 15-120 days and brain tissues were prepared and examined for histological and pathological abnormalities. Based upon this study (Van Dyke, *et al.*, 1987) four grades of tumor pathology could be identified on the basis of tumor size, structure of the tissue, and invasiveness of the neoplasm. Prior to day 36 after birth, no pathological changes were detectable in transgenic mice containing the SV40 large T antigen gene. This was observed in spite of the fact that brain tissue from mice 14-28 days of age expressed detectable levels of the SV40 large T antigen using Western blotting procedures (Van Dyke, *et al.*,

1987). After day 36, mice contained grade I pathology which consisted of multiple microscopic focal areas throughout the choroid plexus (Figure 1). A focus was usually composed of 5-20 cells with darkly staining nuclei, a disorganized and anaplastic appearance, and a higher cell density than normal cells of the choroid plexus. These foci commonly contained mitotic figures which were never observed in normal choroid plexus tissues from mice of comparable ages. Grade II neoplasia occurred in mice older than 42 days and these tumors were larger than grade I (more cells) but did not fill the ventricle nor invade the ependymal lining (Figure 1). Grade III tumors, first observed by 56 days of age, were detectable by macroscopic observation. These tumors retained the well differentiated papillary architecture, filled the ventricle of the brain but were not invasive. Grade IV neoplasms often lost some of the papillary architecture in a portion of the tumor tissue and had areas that invaded the brain parenchyma. The period of most rapid tumor growth usually occurred after 77 days, a time when the levels of SV40 large T antigen increased dramatically in brain tumor tissue (Figure 1 and Table 1) (Van Dyke, *et al.*, 1987).

Table 1

Age (days)	Pathology Grade	Average Tumor size(mm)[a]	Tumor Antigen detected[b]
0-28	0	0	+
28-77	I-III	.29 x .31	+
71-110	III-IV	1.6 x 2.26	+++

a. The sizes of the tumors were averaged from the approximate diameter or two dimensional size.
b. SV40 T antigen was detected in brains or tumors by Western blotting and immunofluorescence of cells in culture (Van Dyke, *et al.*, 1987).

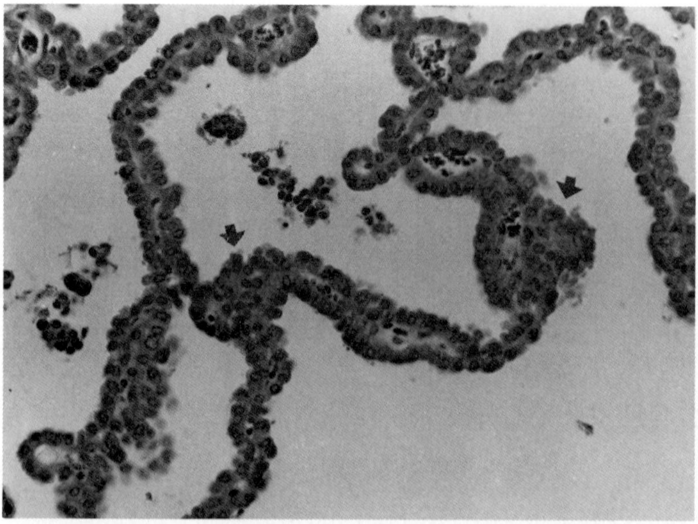

Figure 1A

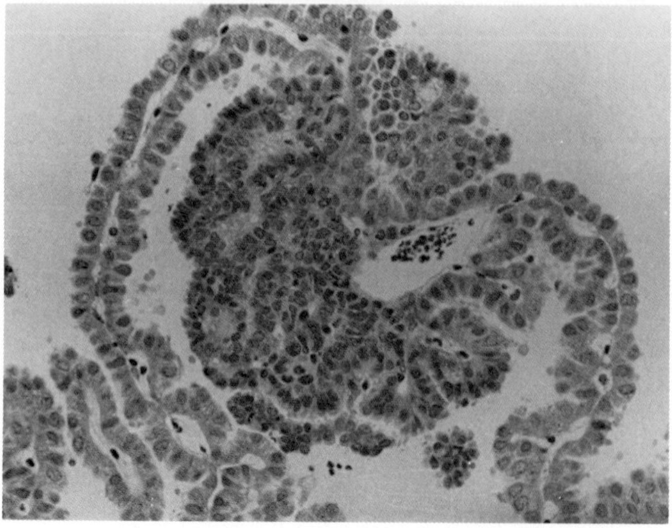

Figure 1B

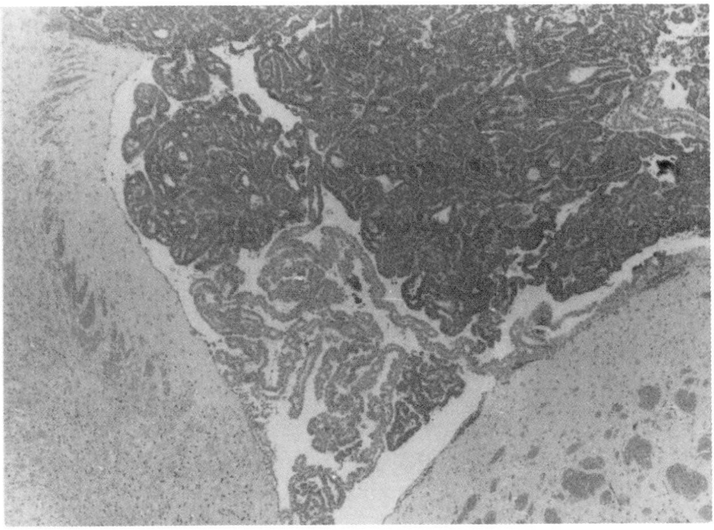

Figure 1C

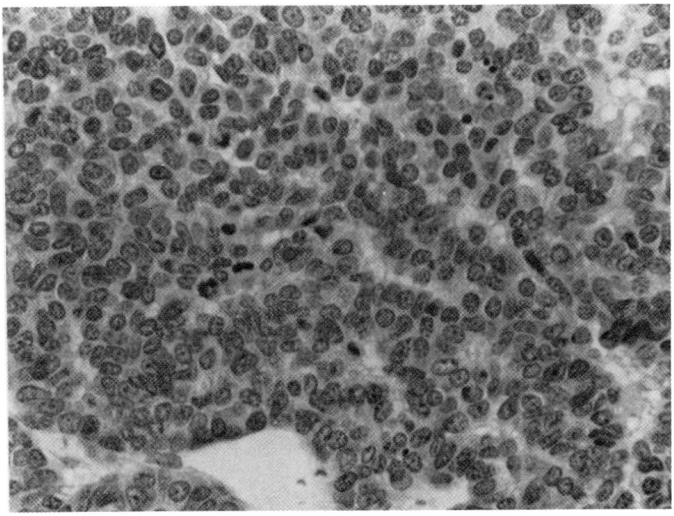

Figure 1D

Figure 1.

The four stages of the choroid plexus neoplasm in SV11 transgenic mice. Brains were fixed by immersion in 4% formaldehyde in PBS and then serially blocked into 0.2cm coronal sections. These sections were embedded in paraffin, sectioned, and stained with a Luxol blue-hematoxylin-eosin combination. A) A photomicrograph of a choroid plexus containing Grade I foci (200X magnification). The arrows point to small foci of cells (5-20) some of which had the characteristics of tumor cells, i.e. darkly staining nuclei, generally disorganized appearance and more crowded than the surrounding choroid plexus. These foci were not macroscopically visible. Mitotic figures were routinely observed in these foci, something which was not seen in normal litter mates or younger SV11 mice with no visible pathology. B) A Grade II tumor (200X magnification). These were also microscopic and still retained papillary architecture characteristic of the choroid plexus. These tumors were much larger than Grade I neoplasms but still did not fill the ventricle or invade the ependymal lining. C) A Grade III neoplasm (100X magnification). These tumors were visible upon macroscopic examination of the sections, often filled the ventricle, but still retained papillary architecture and did not invade the ependyma. D) A Grade IV tumor (400X magnification). These tumors contained areas that were non-papillary and could invade into the parenchyma of the brain.

Based upon these observations, it appears that the SV40 large T antigen is synthesized prior to any detectable pathology, i.e. grade I foci (Table 1). The levels of T antigen increase when rapid tumor growth is detected (after 90 days). In these tumors, T antigen is readily detected by immunoperoxidase staining of tissue sections (Figure 2), while little or no T antigen was detected by this procedure in tissues from younger transgenic mice. (Van Dyke, *et al.*, 1986). These data indicate that multiple foci arise within the choroid plexus epithelial layer of cells several weeks after low levels of T antigen are produced in these cells. Over the

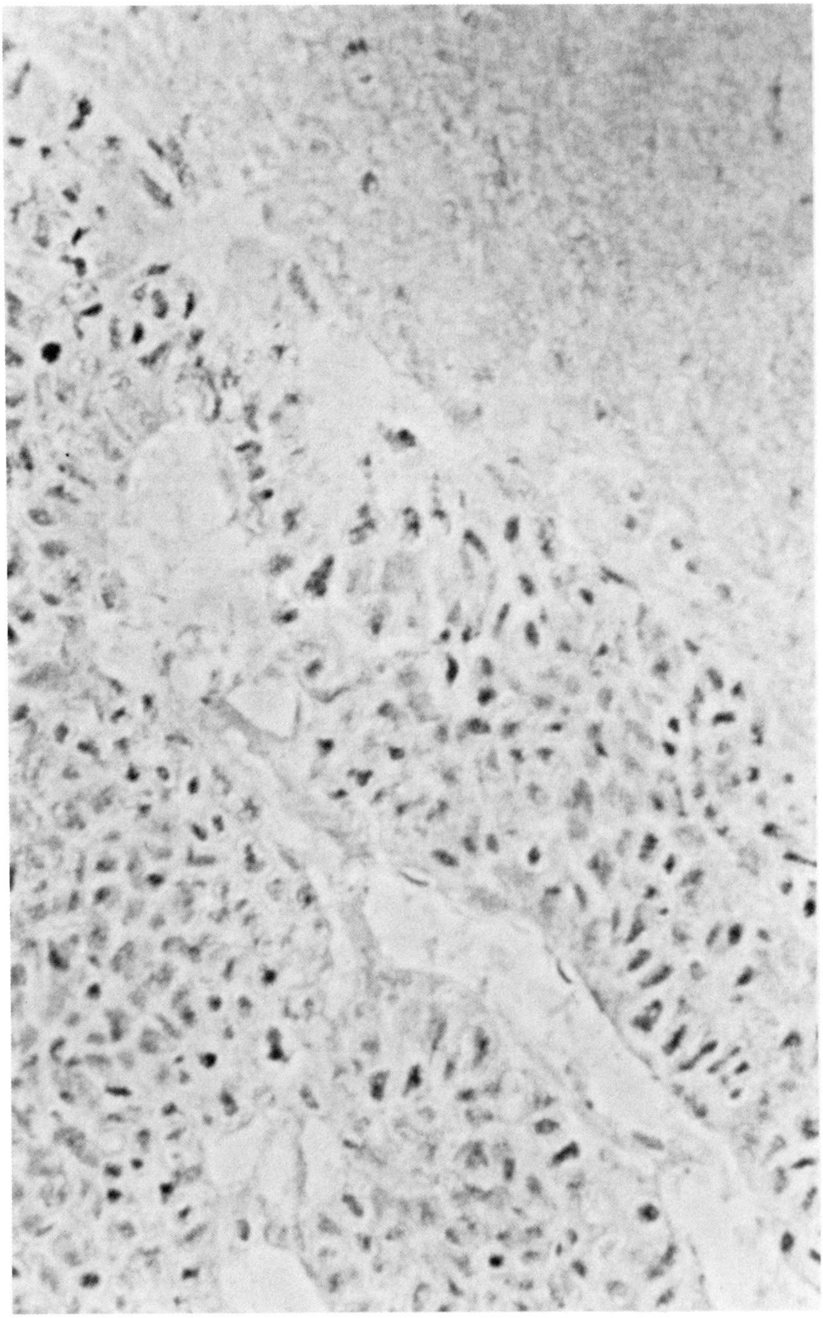

Figure 2

Figure 2.

Immunoperoxidase staining of SV40 large T antigen in a thin section from a tumor bearing brain. Immunohistochemical staining was carried out as described by Van Dyke, *et al.* (1987) using the T antigen monoclonal antibody, PAb412. This photomicrograph shows normal brain next to the positively staining nuclei of the choroid plexus tumor. As expected, T antigen is predominantly localized in the nuclei.

next two month period of life additional events must occur that result in a rapid growth of a focus into a large tumor and higher levels of expression of SV40 large T antigen (by 90 days). Based upon this hypothesis, a set of experiments has been initiated to examine the expression of several cellular genes in the choroid plexus papilloma in an effort to identify possible secondary events that may contribute to the evolution of the tumor in these mice. Both Northern blot hybridization and in situ hybridization with tumor tissue sections have been employed to measure the steady state levels of RNA of several genes. Two classes of cellular genes have been monitored in this way; 1) genes known to preferentially be expressed in the choroid plexus tissue and 2) cellular oncogenes that could be deregulated in such tumors.

MATERIALS AND METHODS

Mice

SV11 mice containing the linearized plasmid, pSV11 (Colby and Shenk, 1982), were maintained as heterozygotes by crossing to normal C57BL6/J mice from Jackson Laboratories. Mice were screened for the presence of the transgene by hybridization to DNA isolated from the tails as described by Brinster, *et al.*, 1984. 419 mice, containing the plasmid pSVMK, were maintained in the same way.

Tumor Cell Culture

Tumor cells were established and maintained in culture as described by Van Dyke, *et al.*, 1985.

RNA Isolation

Tissues were dissected out, flash frozen in liquid nitrogen and stored at -80° until extractions were performed. RNA was extracted by the guanidium/cesium chloride method described by Maniatas, *et al.*, 1982.

Cytoplasmic RNA was extracted from cells in culture by the NP40 lysis method. Briefly, cells were trypsinized and resuspended in 1ml/100mm plate of lysis buffer (30mM Tris pH 7.6, 150mM NaCL, 1.5mM MgCl2, 0.2% NP40 and 10mM vanadyl ribonucleoside complexes (VRC)). Cells were allowed to lyse on ice for 5 minutes. Nuclei were pelleted at 10,000 x g for 5 minutes. The supernatant was spun again and to the final supernatant, an equal volume of proteinase K buffer (100mM Tris pH 7.6, 400mM NaCL, 20mM EDTA, 1% SDS, 10mM VRC) and 100ug/ml proteinase K was added. The solution was incubated at 37° for 20 minutes and then extracted with phenol/chloroform/isoamyl alcohol. The RNA after ethanol precipitation was digested with RNase free DNase (BRL) in 10ml DNase buffer (100mM MgOAC, 200mM Tris pH 7.2), 10ug/ml DNase, and 100 units RNasin (Promega). The reaction was allowed to proceed 30 minutes at room temperature at which time it was stopped by the addition of SDS to 0.5%, EDTA to 50mM and NaOAC to 200mM. The RNA was extracted with phenol/ chloroform/isoamyl alcohol and ethanol precipitated.

Preparation of Probe

Both ^{32}P and ^{35}S labeled DNA probes were prepared by the random priming method described by Feinberg and Vogelstein, 1984. DNA for probe preparation was gel purified using low-melting agarose (FMC) and NACS columns (BRL).

Northern Blot Hybridization

10ug of each RNA was electrophoresed on a formaldehyde agarose gel according to Maniatas, *et al.*, 1982. The gels were stained with ethidium bromide to visualize the ribosomal RNA, photographed, and then blotted to Gene Screen Plus (NEN) according to the supplier's recommendations.

Hybridization was performed as described by Church and Gilbert, 1984. Filters were washed after hybridization twice in 2X SSC at room temperature for 30 minutes followed by two 30 minute incubations in 0.2X SSC, 0.1% SDS at 50°. The filters were then rinsed in 0.1X SSC at room temperature and then wrapped exposed to XAR film (Kodak) while still wet.

In Situ Cytohybridization

Brains were explanted, flash frozen, and stored at -80° until sectioning. To section, brains were cut into 3 parts and embedded in freezing medium. 7 micron sections were cut on a cryostat and thaw mounted onto cleaned slides. These slides were stored unfixed at -80° until hybridized. To hybridize, sections were fixed in 4% paraformaldehyde in PBS at room temperature for 5 minutes. Sections were then washed two times with 2X SSC. Sections were prehybridized with a 50% formamide buffer as described by Shivers, *et al.*, 1986, for 3 hours at room temperature and then hybridized with hybridization buffer (Shivers, *et al.*, 1986) containing 500,000 cpm of a ^{35}S probe in 50 μl in a humidified chamber for 60 hours at room temperature. After hybridization, the sections were washed 2 x 10 minutes in 2X SSC, 0.5% NaPPi at room temperature followed by overnight washing in 0.5X SSC, 0.05% NaPPi, 10mM BMe. The slides were dehydrated in ethanol containing 0.3M NH_4OAc, dried and exposed to XAR film at -80° with an intensifying screen. The slides were then dipped in NTB2 (Kodak) and exposed at room temperature.

RESULTS AND DISCUSSION

I. SV40 T antigen

Previous experiments have demonstrated that SV40 large T antigen expression in transgenic mice is tissue specific (Brinster, *et al.*, 1984) and limited to the choroid plexus (Van Dyke, *et al.*, 1987). In order to confirm and extend those results, Northern blot hybridizations were carried out with RNA preparations isolated from various tissues. RNA prepared from three separate tumors (T2, T3 and T7 in Figure 3A) and a tissue culture cell line prepared from a tumor (TC) all contained SV40 specific RNA sequences about 2 kb in size as expected for the major viral transcript. No detectable SV40 RNA was found in liver ($L3^+$), kidney ($K3^+$) or spleen (data not presented) of SV11 animals containing a tumor (Figure 3A). Brain tissue derived from an SV40 transgenic line of mice (419, Van Dyke, *et al.*, 1986) that failed to produce tumors, did not contain detectable levels of SV40 specific RNA (419 $B1^-$ in Figure 3A).

To localize the SV40 specific RNA within sections of the brain in situ hybridization was carried out. Frozen sections of brain tissue derived from an SV11 mouse with a tumor was hybridized with an ^{35}S-labeled T antigen specific DNA probe. Figure 3B shows macroscopic photographs of sections from two tumors, one from a young mouse (with a small tumor mass) and the second from an older mouse. The autoradiogram presented in Figure 3B demonstrates a preferential localization of SV40 RNA within the tumor mass, with normal brain tissue containing little or no SV40 specific RNA. Disection of the brain into tumor fragments and normal tissue, followed by detecting T antigen by Western blot analysis, also indicates that the SV40 T antigen gene is active preferentially only in tumor tissue. When the bacterial plasmid, without the SV40 sequences, used as a vector was hybridized to thin sections of tumor tissue, no specific hybridization was detected.

The hybridization of an ^{35}S-labeled SV40 specific probe to the tumor tissue was also examined at the microscopic levels using nuclear track emulsion (Figure

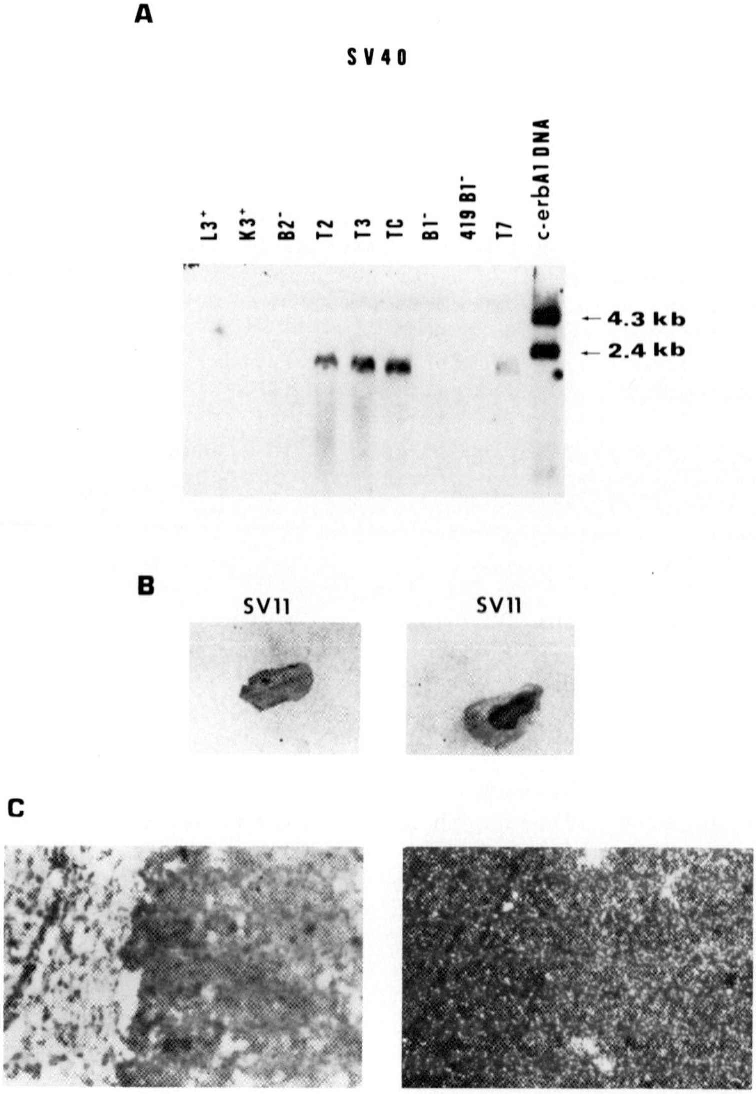
A
SV40
L3+
K3+
B2-
T2
T3
TC
B1-
419 B1-
T7
c-erbA1 DNA
← 4.3 kb
← 2.4 kb
B
SV11
SV11
C

Figure 3

Figure 3.

A) Northern blot of unselected RNA probed with ^{32}P labeled SV40 early region DNA. RNA was from an SV11 liver ($L3^+$), kidney ($K3^+$), normal brain ($B2^-$ and $B1^-$), a brain from a different SV40 transgenic line ($419B1^-$), 3 separate choroid plexus tumors (T2, T3, and T7) and tissue culture passaged tumor cells (TC). B) Autoradiograph of in situ cytohybridization of ^{35}S-labeled SV40 early region DNA to thin sections of SV11 brains. The section on the left shows a small hybridizing nodule in the left lateral ventricle which, upon microscopic examination, proved to be a tumor mass. C) Photomicrographs of in situ cytohybridization as in B exposed to nuclear track emulsion for 10 days. Bright and dark field microscopy shows the silver grains as dark and bright spots respectively.

3C). The cellular border between tumor and normal brain is shown in these photomicrographs. Significantly higher grain density is apparent over cells in the tumor mass with some cells, or groups of cells, showing a much higher grain density. Both bright field and dark field photomicrographs are presented. In the dark field view, the bright spots are the silver grains. These sections were cut at a thickness of seven microns which should be an average of one cell thick. One cannot eliminate the possibility that the closer packing of cells or the piling up of cells in the tumor results in a higher localized grain density. It is more likely that selected cells in the tumor have higher levels of T antigen specific mRNA. Given the observation that tumor tissue from older animals has higher levels of T antigen and grows more rapidly (after 90 days), the heterogeneity of T antigen mRNA levels in a population of tumor cells might allow for selection of cells with high T antigen levels and rapid growth.

II. Differentiated Gene Functions of the Choroid Plexus

The choroid plexus is a highly specialized structure in the brain responsible for the elaboration of the cerebrospinal fluid (CSF) derived from the plasma. The

CSF, however, is not simply a filtrate of the blood proteins but its composition is selectively enriched and depleted of various components (Davson, 1967; Miner and Reed, 1972). Several genes are selectively expressed in the choroid plexus epithelial cell layer. Transthyretin (TTR), which participates in the transport of retinol and thyroxine in the plasma, is synthesized primarily in the liver and the choroid plexus (Soprano, *et al.*, 1985). TTR is found in very high concentrations in the CSF. In human CSF, TTR can contribute up to 25% of the protein suggesting the choroid plexus produces or transports a great deal of this protein (Weisner, *et al.*, 1983). Employing a 500 base pair probe for TTR specific nucleotide sequences, a variety of RNAs from transgenic mice were examined for this gene product. The Northern blot hybridizations, (Figure 4A) with RNAs from the liver ($L1^+$), brains with tumors (BWT1, 2) or from isolated tumors (T11) of the choroid plexus of SV11 transgenic mice, detected a 0.7 kb RNA of the correct size previously reported for the transthyretin mRNA (Soprano, *et al.*, 1985; Herbert, *et al.*, 1986). In liver tissues from C57BL6/J mice (not transgenic for SV40 DNA), who were litter mates of the SV11 mice, TTR specific RNA was also detected ($L1^-$, Figure 4A). In longer exposures of the autoradiogram, the RNA from normal brain tissue ($B1^-$) also shows a 0.7 kb TTR specific RNA species. RNA from kidney or spleen tissues (K^-, $S1^+$) contained no detectable mRNA hybridizing with the TTR probe.

In situ hybridization to sections of normal brain or brain sections with papillomas of the choroid plexus (Figure 4B) showed ventricular regions in normal brains and obvious tumors in the SV11 brain sections hybridizing intensely. The normal tissue section shows the two lateral ventricles hybridizing with this probe (the third ventricle is not present in this section). The tumor taken from a young mouse (Figure 4B, middle section) indicates a small tumor enlargement in both lateral ventricles (two independent tumors) and a normal sized third ventricle. Of 19 tumors examined histologically (Van Dyke, *et al.*, 1987) a neoplasia of the third ventricle has never been observed. In older SV11 mice the tumor progresses rapidly and can occupy almost half the thin section (Figure 4B). The intensity of TTR specific grain density in sections of normal tissue, young tumors

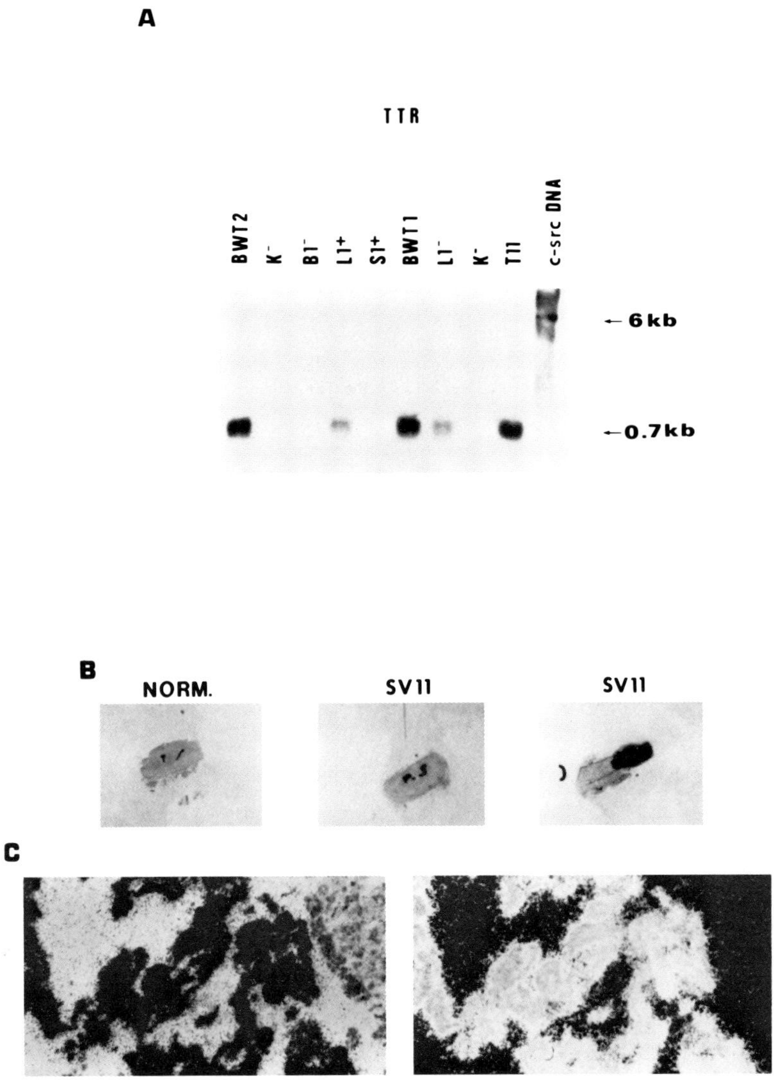
A
TTR
BWT2
K-
B1-
L1+
S1+
BWT1
L1-
K-
T11
c-src DNA
← 6kb
← 0.7kb
B
NORM.
SV11
SV11
C

Figure 4

Figure 4.

A) Northern blot, as in Figure 3, probed with a labeled transthyretin cDNA clone. RNA was from normal kidneys (K^-), normal liver (L^-), normal brain ($B1^-$), SV11 liver ($L1^+$), spleen ($S1^+$), choroid plexus tumor (T11), and 2 separate brains containing tumors (BWT). B) In situ hybridization of thin sections hybridized with ^{35}S-labeled TTR. A normal brain and 2 stages of choroid plexus tumors are shown. C) Bright and dark field photomicrographs of *in situ* hybridization. Normal brain is in the upper right hand corner of the photographs and the empty area is ventricular space.

or advanced tumors appears at this level of resolution to be about equal. The microscopic bright and dark field photos of a tumor section (Figure 4C) show a high grain density distributed evenly over all cells (nearly obscured by the grains) in the section. Little or no TTR specific hybridization was observed outside the tumor. These experiments demonstrate that the expression of TTR transcripts is not altered by the neoplasia of the choroid plexus.

A second tissue restricted gene product preferentially found in the choroid plexus is the serotonin receptor (Yagaloff, *et al.*, 1986). While many different serotonin binding sites have been characterized, the specific receptor that binds serotonin and its agonists in the choroid plexus is known as the 5-HT_{1C} receptor (Yagaloff and Hartig, 1985). A portion of the cDNA for this receptor has been cloned from RNA derived from these choroid plexus tumors of SV11 mice (Lubbert, *et al.*, 1987). Sections of tumor tissue express high levels of the 5-HT_{1C} receptor (Yagaloff, *et al.*, 1986). To examine the RNA levels of the 5-HT_{1C} receptor in tissues from these transgenic mice Northern blot hybridization was employed (Figure 5). A 6 kb transcript was readily detected in three separate tumors of the choroid plexus (T1, T2, and T3) and faint hybridization to RNA from normal brain tissue was also observed ($B2^-$, $B1^-$). Cells derived from these papillomas of the choroid plexus and grown in cell culture rapidly lose the ability to bind serotonin agonists (P. Hartig, personal communication).

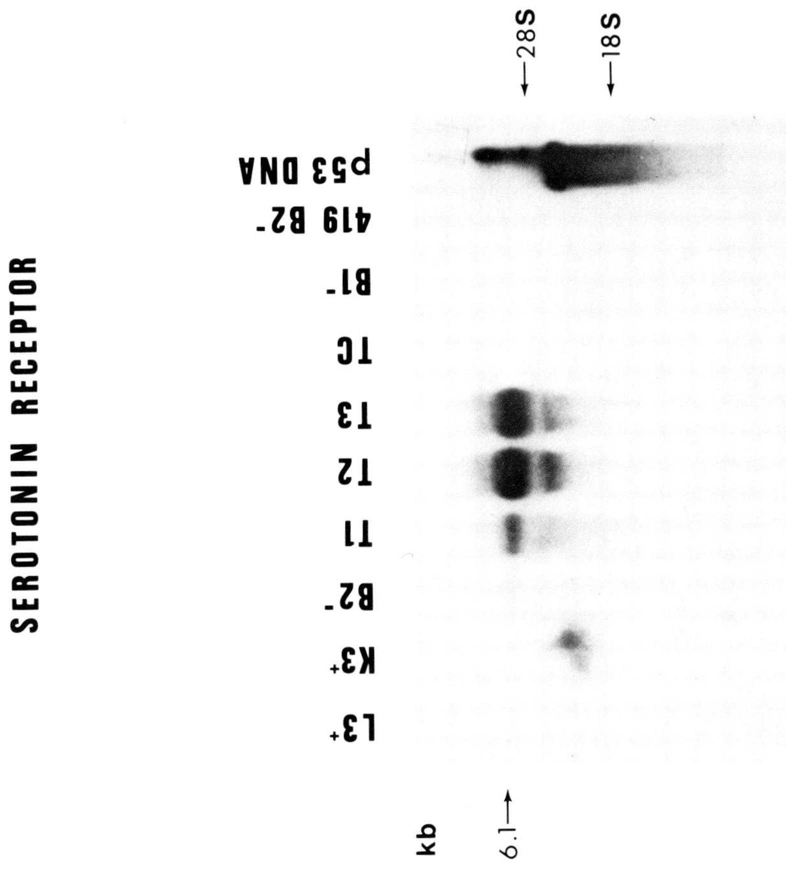
SEROTONIN RECEPTOR
L3+
K3+
B2-
T1
T2
T3
TC
B1-
419 B2-
p53 DNA
28S
18S
kb
6.1

Figure 5

Figure 5.
A Northern blot, as in Figure 3, probed with a labeled 5-HT_{1C} cDNA clone. RNA was from normal brain ($B1^-$ and $B2^-$), a brain from another SV40 transgenic line ($419B2^-$), SV11 liver ($L3^+$), kidney ($K3^+$), 3 separate tumors (T1, T2, and T3), and tissue culture passaged tumor cells (TC).

RNA extracted from tumor cells maintained in culture (Figure 5, TC) no longer contains the serotonin receptor transcript. These cells then lose the receptor due to a lack of expression of this gene.

The expression of both the TTR and the 5-HT_{1C} receptor mRNAs in tumor cells demonstrates that these tumors retain several differentiated functions. This is consistent with the normal papillary architecture of choroid plexus tumors (Figure 1).

III. p53-oncogene expression in choroid plexus tumors.

p53 is a cellular protein found in normal cells at very low levels and transformed cells or tumors at much higher levels (Crawford, 1983; Thomas, *et al.*, 1983). In nontransformed cells p53 has a short half-life (20 minutes) while in transformed cells p53 protein has a longer half-life resulting in higher concentrations of this protein (Oren, *et al.*, 1981). In SV40 transformed cells p53 is found in an oligomeric protein complex with the SV40 large T antigen (Lane and Crawford, 1979; Linzer and Levine, 1979) which results in a prolonged p53 half-life and high levels of p53 protein (Oren, *et al.*, 1981). Increased levels of p53 mRNA have been detected in several normal tissues (such as spleen) and tumor tissues (Oren, *et al.*, 1982; Reich, *et al.*, 1983; Rogel, *et al.*, 1985). Clearly, both protein stability and mRNA levels may regulate the amount of p53 protein in transformed cells (Reich, *et al.*, 1983). The fact that the SV40 large T antigen can regulate the level of p53 in a transformed cell (Linzer, *et al.*, 1979) and that p53 cDNA and genomic clones have transforming activities (Eliyahu, *et al.*, 1984; Parada, *et al.*, 1984) suggested that the levels of p53 specific RNAs should be examined in the choroid plexus tumors of SV11 mice. Northern blot hybridization using a p53 cDNA probe directed against RNA from a variety of tissues detected a 2 kb message in liver ($L3^+$), kidney ($K2^+$, $K3^+$), spleen ($S2^+$) and tumor tissue (T4, T5, T6, and T9) of transgenic mice (Figure 6). The spleen contained very high levels of p53 mRNA and three of the four tumor tissues contained comparable levels (T5, T6, and T9) while one tumor had lower levels of p53 RNA. Brain tissue from a transgenic mouse containing the SV40 genome that failed to express SV40 T antigen RNA (419 $B3^-$) had much reduced levels of p53 RNA (Figure 6). In situ hybridization of the p53 probe with sections of normal or tumor tissue (Figure 6) shows the localization of p53 specific RNA to the tumors of the choroid plexus. Normal choroid plexus tissue failed to localize the p53 probe in any region of the brain section while the small tumor from a young animal or the large tumor from an older animal (Figure 6B) localized the labeled p53 probe in the tumor of the choroid plexus. The high magnification photomicrographs show that the grain density over cells in the tumor (Figure 6C) was much greater than over the choroid plexus from normal brain. The grain density with the p53 probe in sections of normal tissue was evenly distributed over choroid plexus tissue, ventricular tissue and non-choroidal brain cells. In sections from a tumor bearing brain, the border between tumor and normal cells is demarcated by a grain density change, with tumor cells having most of the p53 probe associated with them.

The steady state levels of p53 mRNA and protein are known to increase about five fold after the entrance of quiescent cells into the cell cycle (Reich and Levine, 1984; Mercer and Baserga, 1985). This could explain the increased levels of p53 RNA found in the rapidly dividing tumor cells. Other cellular oncogenes which are known to increase their levels of mRNA upon entry into the cell cycle, such as myc and fos (Kelly, *et al.*, 1983; Greenberg and Ziff, 1984) did not exhibit consistently increased levels of RNA in these choroid plexus tumors. Both Northern blot analysis and in situ hybridizations failed to show comparable increases in myc or fos RNA levels in these tumors as was observed with the p53 probe.

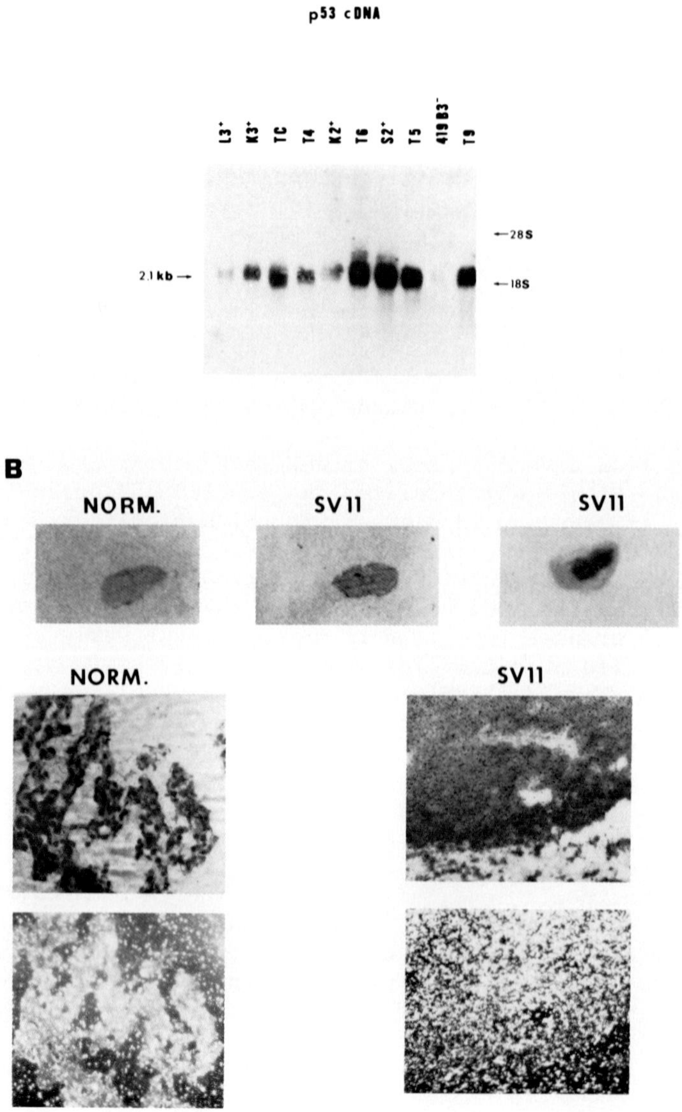
p53 cDNA
L3+
K3+
TC
T4
K2+
T6
S2+
T5
419B3-
T9
2.1 kb
28S
18S
B
NORM.
SV11
SV11
C
NORM.
SV11

Figure 6

Figure 6.
A) Northern blot, as in Figure 3, hybridized with a labeled p53 cDNA clone. RNA was from SV11 liver ($L3^+$), kidney ($K3^+$ and $K2^+$), spleen ($S2^+$), a brain from another SV40 transgenic line ($419B3^-$), 4 separate tumors (T4, T5, T6, and T9) and tissue culture passaged tumor cells (TC). B) Autoradiographs as in Figure 3 of p53 in situ hybridized brain sections. The SV11 brain sections are from the same brains as in Figure 3. Note the co-localization of hybridization with p53 and SV40 probes. C) Photomicrographs of normal choroid plexus and choroid plexus tumor in situ hybridized with p53 and exposed to nuclear track emulsion with bright field on the top and dark field below.

While these data demonstrate an enhanced level of p53 specific RNA in choroid plexus tumors, it remains possible that this is simply a reflection of the increased rate of cell division. Alternatively, the expression of T antigen or the tumorigenic state of these cells, could activate higher levels of p53 mRNA. The appropriate control for choosing between these alternatives will be to examine normal dividing choroid plexus tissue which is found at a low frequency after birth (Chauhan and Lewis, 1979). This will have to be done because it is not even clear that cell cycle regulated events described in fibroblasts in cell culture are applicable to epithelial cells of the choroid plexus.

References

Brinster RL, Chen HY, Messing A. Van Dyke T, Levine AJ, Palmiter R (1984) Transgenic mice harboring SV40 T antigen genes develop characteristic brain tumors. *Cell* 37:367-379.

Chauhan AN, Lewis PD (1979) A quantitative study of cell proliferation in ependyma and choroid plexus in the post-natal rat brain. *Neuropath and App. Neurobiol.* 5:303-309.

Church GM, Gilbert W (1984) Genomic sequencing. *PNAS* 81:1991-1995.

Colby WW, Shenk T (1982) Fragments of the simian virus 40 transforming gene facilitate transformation of rat embryo cells. *PNAS* 79:5189-5193.

Crawford LV (1983) The 53,000-dalton cellular protein and its role in transformation. *Int. Rev. Exp. Pathol.* 25:1-50.

Davson H (1967) "Physiology of the cerebrospinal fluid." Little, Brown and Company.

Eliyahu D, Rax A, Gross P, Givol D, Oren M (1984) Participation of p53 cellular tumour antigen in transformation of normal embryonic cells. *Nature* 312:646-649.

Feinberg AP, Vogelstein B (1984) A technique for radiolabeling DNA restriction endonuclease fragments to high specific activity. *Anal. Biochem.* 137:266-267

Greenberg ME, Ziff EB (1984) Stimulation of 3T3 cells induces transcription of the c-fos proto-oncogene. *Nature* 311:433-442.

Herbert J, Wilcox JN, Pham KTC, Fremeau RT, Zeviani M, Dwork A, Soprano DR, Makover A, Goodman DS, Zimmerman EA, Roberts JL, Schon EA (1986) Transthyretin: a choroid plexus-specific transport protein in human brain. *Neurol.* 36:900-911.

Kelly K, Cochran BH, Stiles CD, Leder P (1983) Cell-specific regulation of the c-myc gene by lymphocyte mitogens and platelet-derived growth factor. *Cell* 35:603-610.

Lane DP, Crawford LV (1979) T antigen is bound to a host protein in SV40-transformed cells. *Nature* 278:261-263.

Linzer DIH, Levine AJ (1979) Characterization of a 54K dalton cellular SV40 tumor antigen in SV40 transformed cells. *Cell* 17:43-51.

Lübbert H, Hoffman BJ, Snutch TP, Van Dyke T, Levine AJ, Hartig PR, Lester HA, Davidson N (1987) cDNA cloning of a serotonin 5-HT_{1C} receptor using electrophysiological assays of mRNA injected xenopus oocytes. *PNAS* 84:4332-4336.

Maniatas T, Fritsch EF, Sambrook J (1982) "Molecular cloning." Cold Spring Harbor Laboratory.

Mercer WE, Baserga R (1985) Expression of the p53 protein during the cell cycle of human peripheral blood lymphocytes. *Exp. Cell Res.* 160:31-46.

Miner LC, Reed DJ (1972) Composition of fluid obtained from choroid plexus tissue isolated in a chamber in situ. *J. Physiol.* 227:127-139.

Oren M, Maltzman W, Levine AJ (1981) Post translational regulation of the 54K cellular tumor antigen in normal and transformed cells. *Mol. Cell. Biol.* 1:101-110.

Palmiter R, Chen H, Messing A, Brinster R (1985) SV40 enhancer and large T antigen are instrumental in development of choroid plexus tumors in transgenic mice. *Nature* 316:457-460.

Parada LF, Land H, Weinberg RA, Wolf D, Rotter V (1984) Cooperation between genes encoding p53 tumour antigen and *ras* in cellular transformation. *Nature* 312:649-651.

Reich NC, Levine AJ (1984) Growth regulation of a cellular tumor antigen, p53, in non-transformed cells. *Nature* 308:199-201.

Reich NC, Oren M, Levine AJ (1983) Two distinct mechanisms regulate the levels of a cellular tumor antigen, p53. *Mol. Cell Biol.* 3:2143-2150.

Rogel A, Popliker M, Webb CG, Oren M (1985) p53 cellular tumor antigen: analysis of mRNA levels of normal adult tissues, embryos and tumors. *Mol. Cell. Biol.* 5:2851-2855.

Shivers BD, Schacter BS, Pfaff DW (1986) In situ hybridization for the study of gene expression in the brain. ed. Conn PM: "Methods in Enzymology" Academic Press, Vol. 124.

Soprano DR, Herbert J, Soprano KJ, Schon EA, Goodman DS (1985) Demonstration of transthyretin mRNA in the brain and other extrahepatic tissues in the rat. *J. Biol. Chem.* 260:11793-11798.

Thomas R, Kaplan L, Reich N, Lane DP, Levine AJ (1983) Characterization of human p53 antigens employing primate specific monoclonal antibodies. *Virology* 131:502-517.

Van Dyke T, Finlay C, Levine AJ (1986) A comparison of several lines of transgenic mice containing SV early genes. *Cold Spring Harbor Symp. Quant. Biol.* 50:671-678.

Van Dyke TA, Finlay C, Miller D, Marks J, Lozano G, Levine AJ (1987) Relationship between simian virus 40 large tumor antigen expresson and tumor formation in transgenic mice. *J. Viol.* 61:2029-2032.

Weisner B, Roethig H-J (1983) The concentration of prealbumin in cerebrospinal fluid, indicator of CSF circulation disorders. *Eur. Neurol.* 22:96-105.

Yagaloff KA, Hartig PR (1985) ^{125}I-lysergic acid diethylamide bonds to a novel serotonergic site on rat choroid plexus epithelial cells. *J. Neurosci.* 5:3178-3183.

Yagaloff KA, Lozano G, Van Dyke T, Levine AJ, Hartig PR (1986) Serotonin 5-$HT1_C$ receptors are expressed at high density on choroid plexus tumors from transgenic mice. *Brain Res.* 385:389-394.

Cellular Factors in Development and Differentiation:
Embryos, Teratocarcinomas, and Differentiated Tissues, pages 187–209

Involvement of Fos as a Trans-Acting Factor in Adipogenic Gene Expression

Robert J. Distel and Bruce M. Spiegelman

Dana-Farber Cancer Institute and Department of Biological Chemistry and Molecular Pharmacology, Harvard Medical School, 44 Binney St., Boston, MA 02115

INTRODUCTION

Cell differentiation is a complex process which involves changes in cell growth, morphology, surface interactions between cells and the production of specialized proteins. In general all of these processes have been shown to be influenced or regulated by changes in the pattern of gene expression. Since transcriptional control is the most common mechanism implicated for gene regulation, we are led to two difficult questions: (1) what are the mechanisms that regulate the transcription of a particular gene and (2) what is the circuitry whereby extracellular signals activate the correct program of gene expression?

To study these probems in the adipose cell lineage we and others have cloned several genes that are characteristic of mature adipocytes (Spiegelman et al., 1983; Bernlohr et al., 1984; Chapman et al., 1984; Doglio, 1986). This was made possible by the establishment of preadipocyte mouse cell lines which can differentiate into adipocytes that resemble their in vivo counterparts both morphologically and biochemically (Green, 1979; Sidhu, 1979; Spiegelman and Green, 1979).

We have concentrated our studies on three genes, all of which are expressed in adipocytes and undergo differentiation-dependent increases in transcription (Cook et al., 1985a; Bernlohr et al., 1985; Djian et al., 1985; Doglio et al., 1986). These include glycerophosphate dehydrogenase (GPD), an enzyme crucial for triglyceride metabolism, and two previously unidentified and apparently novel gene products: adipsin, a serine protease homologue specifically secreted by fat cells (Spiegelman et al., 1983; Cook et al., 1985b; Zezulak et al., 1985; Cook et al., 1987) and adipocyte P2 (aP2) a protein of 13 Kd with strong homology to myelin P2 and more limited homology to a broad family of intracellular lipid-binding proteins (Spiegelman et al., 1983; Bernlohr et al., 1984; Cook et al., 1985a).

The isolation of genomic clones containing 5' and 3' flanking regions for these genes (Hunt et al., 1986; Phillips et al., 1986) has made it possible to search for common sequence elements which may play a role in differentiation-dependent transcription. Computer-assisted comparison of the three genes revealed two groups of homologies we have designated fat specific elements (FSE) 1 and 2 (Hunt et al., 1986; Phillips et al., 1986). FSE1 is a 13 bp sequence with a consensus sequence GGC T/A CTGGTCA G/T G and is found in multiple copies upstream of all three genes. aP2 and GPD contain a 13 bp sequence designated FSE2, which differ only by the addition of a single guanidine in GPD (Figure 1). The adipsin gene contains a 7 base region of homology. The presence of common sequence elements in the promoter regions of these three genes which code for members of different protein families suggests a possible regulatory function. This possibility was bolstered by homologies of these sequences to regulatory regions that had been identified in other genes. The last 7 bases of the FSE1 sequence closely matches the 5' end of the BK virus enhancer core (Rosenthal et al., 1983), whereas the FSE2 sequence resembles both the putative enhancer core element of the polyoma virus (Weber et al., 1984) and the negative regulatory element of the human beta-inteferon IRE region (Goodbourn et al., 1986). In light of the finding that a number of regulatory elements in other eucaryotic genes have been shown to bind trans-acting factors, we examined the FSE2 sequence for its ability to bind nuclear proteins in a sequence specific fashion from preadipocytes and adipocytes. We show here that the aP2-FSE2 site binds a protein complex that is altered during cell differentiation and functions in the control of aP2 gene expression. In addition, using immunological approaches we have identified the proto-oncogene product c-fos as a component of the complex that is directly required for sequence-specific binding.

RESULTS

Sequence-Specific Binding of Adipocyte Nuclear Factors

Complementary synthetic oligomers were constructed for the aP2-FSE2 sequence consisting of the 13 bp core homology and 9 flanking bp with overhanding Sal1 and Xho1 sites on either end (Figure 1). Radiolabeled aP2-FSE2 oligomers were incubated with nuclear extracts from 3T3-F442A adipocytes as described by Distel et al. (1987), and protein-DNA complexes were separated from free DNA by electrophoresis on low ionic strength polyacrylamide gels (Fried and Crothers, 1981; Garner and Revzin, 1981). The FSE2 oligomer binds nuclear factors from adipocytes as demonstrated by its retardation in mobility relative to the free oligomer at the bottom of the gel (Figure 2). The specificity of the binding was determined by forming the complexes in the presence of specific and nonspecific competitors. Figure 2 shows that while 10 ng of unlabeled homologous oligomer reduced the retarded band almost completely, similar mounts of a 28 bp oligomer made to the aP2-FSE1 sequence (Hunt et al., 1986), a heterologous 37 bp oligomer, N106, from the bovine rhodopsin gene, double stranded poly(dI:dC), and salmon sperm DNA all are relatively ineffective at reducing the binding.

ALIGNMENT OF FSE2 ELEMENTS

```
                   -178
Adipsin              GCAGGAGGGAGAGGAGGGTTTGC
                   -188          |||||||
GPD FSE2             GGGAACTCGAGAGGAAAAGCTGG
                    -122    \\\\||||||||||
aP2 FSE2              CATGACTCAGAGGAAAACATAC

     -160                -140                -120                -100
  TGATCATTGCCAGGGAGAACCAAAGTTGAGAAATTTCTATTAAAAACATGACTCAGAGGAAAACATACA
                 -80                 -60                 -40
GGGTCTGGTCATGAAGGAAATGATCTGGCCCCCATTGGTCACTCCTACAGTCACATGGTCAGGGCATCTTT
     -20                 +1                  +20
AAAAGTGAGCTATCTGGACTTCAGAGGCTCATAGCACCCTCCTGTGCTGCA
```

Figure 1. The aP2 Gene 5'-Flanking Region and FSE-like Sequences. As reported by Hunt et al. (1986), the aP2 FSE2 sequence is at -118, FSE1 is at -99, an inverted CAAT box is at -66, and a TATA box is at -31. Above the aP2 5'-flanking sequence is a series of FSE2 and FSE2-like sequences derived from other genes that were used to make synthetic oligomers. These include aP2 FSE2, the GPD FSE2, and the adipsin FSE2-like sequence (Hunt et al., 1986; Phillips et al., 1986). The BclI-Mbol restriction fragment (-168 to -79) was used for DNA retardation gels.

Sequence specificity was also examined on an 89 bp Bcl1-Mbo1 restriction fragment of the aP2 gene (-168 to -79 in Figure 1). This restriction fragment contains both the FSE2 sequence and a member of the FSE1 family but falls short of the inverted CCAAT element at -66 (Figure 1). In promoter fusion experiments, we (Distel et al., 1987) have demonstrated that constructions containing from -168 to +21 of the aP2 gene linked to the bacterial chloramphenicol acetyltransferase gene are sufficient to give differentiation-dependent regulation. Thus, the 89 bp restriction fragment contains most of the

DNA 5' to the CCAAT element that regulates differentiation-dependent transcription. Competition with the aP2-FSE2 oligomer virtually eliminates the retarded bands derived from the 89 bp restriction fragment, suggesting that the major binding protein recognizing this DNA is bound at or near the FSE2 site (Figure 2B). Further evidence comes from DNase1 protection experiments using the 89 bp restriction fragment, which gave a protected region over the FSE2 element (Distel et al., 1987). The FSE1 sequences, which were identified by comparison of the same three genes, appear not to bind factors under these conditions (Distel et al., 1987) and do not compete for the FSE2-binding factors. It appears then that the specific binding to the FSE2 site is the major detectable binding site upstream of the canonical CCAAT element on this fragment.

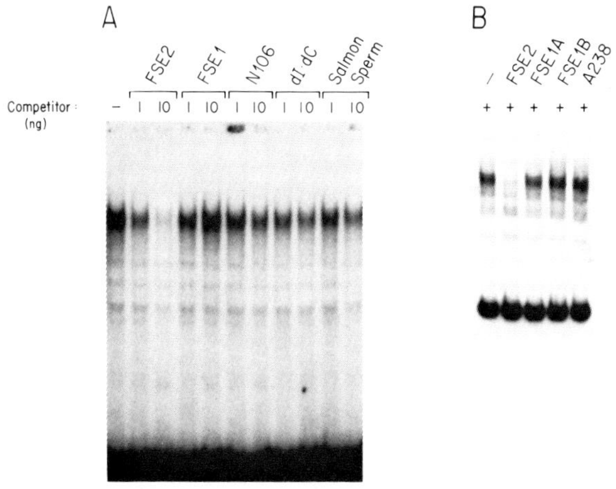

Figure 2. The aP2 FSE2 Sequence Binds Nuclear Factors from Adipocyte Nuclear Extracts in a Sequence-Dependent Fashion. Approximately 0.1 ng of ^{32}P-end-labeled, 28 bp aP2 FSE2 synthetic oligonucleotide (A) or the 89 base BclI-Mbol fragment from the aP2 gene (B) was incubated with adipocyte nuclear extracts as described in Distel et al. (1987). One or 10 ng (A) or 10 ng (B) of competitor DNAs (see Distel et al., 1987)

was added simultaneously with the labeled sequence.

Altered Patterns of FSE2 Binding During Fat Cell Differentiation.

The increase in transcription of the aP2 gene during the differentiation of these cells led us to examine whether the binding pattern of the FSE2 sequence changed from preadipocytes to adipocytes. Preadipocyte nuclear extract was incubated with both the aP2 FSE2 sequence and the 89 bp restriction fragment. Preadipocyte nuclear extract contains factors that bind the aP2 FSE2 oligomer, producing a smear rather than a band as seen in adipocytes (Figure 3A). The smear migrates slightly faster than the adipocyte band, traveling one-third to halfway down the gel, indicating that preadipocyte complexes differ from the adipocyte complexes. Competitions with other oligomers indicates that this smear is sequence specific (Figure 3A).

The binding pattern of preadipocyte nuclear extracts on the 89 bp restriction fragment is somewhat more complex. It gives a smear that runs faster than the band from adipocyte extract and is competed away only by the FSE2 oligomer, indicating that the smear described previously is not a function of the size of the oligomer (Figure 3B). In addition, the preadipocyte extract contains a strong band that is not competed by the FSE2 or the FSE1 oligomers but is specifically competed by a 190 bp Bcl1-Pst1 fragment that contain the 89 bp fragment. This suggest that unlike the adipocyte extract, preadipocyte extracts contain at least two factors that recognize this fragment: one which binds at the FSE2 site and accounts for the smear, and one that binds elsewhere and gives rise to the band.

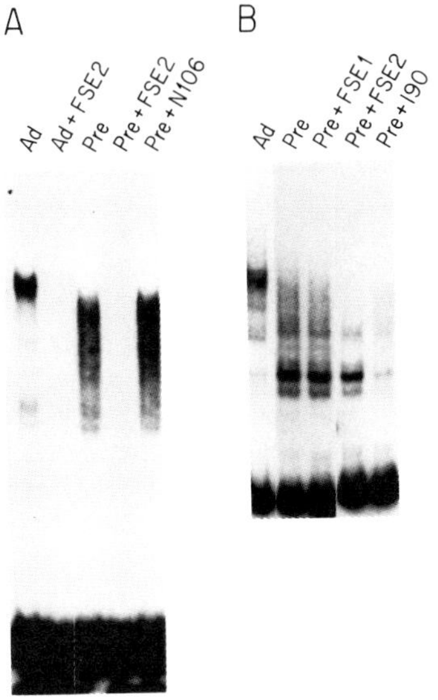

Figure 3. Altered FSE2-Binding Factors during Fat Cell Differentiation. Binding assay with nuclear extracts was done as described in Distel et al. (1987). Ad and Pre are extracts from adipocytes and preadipocytes incubated with the aP2 FSE2 synthetic oligonucleotide (A) or the 89 bp Bcl1 Mbo1 fragment from the aP2 gene. (+) addition of cold competitor DNAs at 10 ng; (190) the 190 bp Bcl1-Pst1 fragment as shown in Figure 1.

Characterization of the Changes in FSE2 Binding During Differentiation.

In transfection assays, the FSE2 site appears to function as a negative regulator of aP2 expression in preadipose cells but not in fat cells (Distel et al., 1987). Because of this and the striking change in mobility of the complexes during differentiation, it is important to understand the differentiation-dependent structural changes at this site. We have considered two possibilities for the change from a smear to a band during differentiation. One possibility was that the half-life of the DNA-protein complex from preadipocytes was much shorter than that from adipocytes, resulting in the dissociation of the preadipocyte complexes during electrophoresis and the generation of free (faster moving) DNA. The other possibility was that the preadipocyte smear was the result of a large number of stable but heterogeneous nucleoprotein complexes. To first address these possibilities, the half-time of dissociation of preadipocyte and adipocyte complexes in solution was estimated. Complexes from preadipocytes and adipocytes were formed as described, and the reaction was stopped by the addition of EDTA and excess cold FSE2 DNA. The complexes were run on nondenaturing gels at 0, 10, 30, 60 and 120 minutes after the complexes formation was stopped. Figure 4 demonstrates that the adipocyte and preadipocyte complexes have similar half-lives (about 2 hours), and the preadipocyte pattern does not change in complexity as a function of dissociation time. In the second experiment, we excised different parts of the preadipocyte pattern and asked whether they electrophoresed as they did originally. As shown in Figure 5, the second electrophoresis run produced bands which ran at their original positions, indicating that the preadipocyte smear is actually a very heterogeneous mixture of discrete and stable complexes and is not caused by nucleoprotein dissociation during electrophoresis.

Direct Participation of c-fos in the aP2 FSE2 Nucleoprotein Complex

We have sought to learn the identity of the protein(s) constituting the FSE2 complex. Because nuclear oncogenes such as c-fos have been implicated in the transduction of extracellular signals related to cell growth and differentiation (Greenberg and Ziff, 1984; Kruijer et al., 1984; Muller et al., 1984; Morgan and Curran, 1986), we examined the effects of antibodies directed against c-fos and some other nuclear oncogenes on the FSE2 complexs. Antibodies were added to adipocyte and preadipocyte nuclear extracts and incubated for 16 hours at 4°C before the addition of

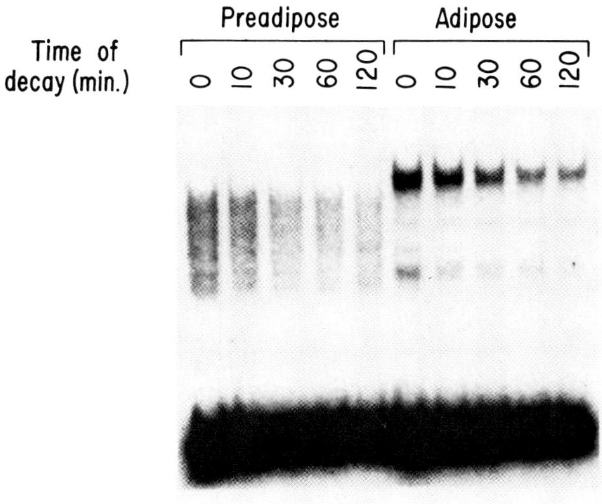

Figure 4. Dissociation Rates of the FSE2 DNA-Protein Complexes. DNA protein complexes were formed with nuclear extract from preadipocytes and adipocytes on the aP2 FSE2 oligomer as described in Distel et al. (1987). The complexes were incubated for the times shown after the addition of 5 mM EDTA and 40 ng of cold aP2 FSE2 oligomer and then electrophoresed on nondenaturing gels.

labeled aP2 FSE2 DNA. Figure 6 indicates that the addition of affinity purified anti-c-fos antibody (M-peptide antibody; Curran et al., 1984) prevents formation of the normal adipocyte protein complex and generates some counts in a slower migrating band. The same decrease is seen in the preadipocyte pattern with a much more pronounced conversion to the slow migrating form. Other purified antibodies including anti-c-myc, anti-SV40 large T-antigen and anti-actin have no detectable effect on the sequence-specific pattern. A nonspecific band seen in both adipocyte and preadipocyte patterns was unchanged by this antibody.

To determine whether this blocking was actually due to antigen recognition by the anti-c-fos antibody, this preparation was preincubated with and without excess M-peptide

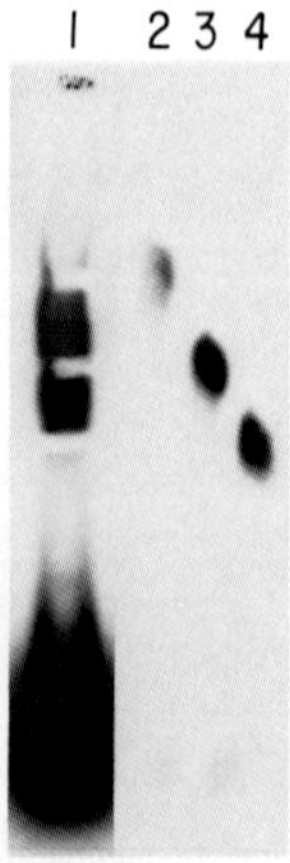

Figure 5. Preadipose FSE2 Pattern Consists of a Mixture of Discrete Complexes. Preadipose DNA-protein complexes were made as in Figure 3 but scaled up 10-fold. The complex was run on a nondenaturing gel for 60 minutes, three bands were excised from lane 1 and electrophoresed on a second nondenaturing gel for 60 minutes (Lanes 2-4). The bands rerun at their original positions.

used to raise this antibody, and the mixture was added to the nuclear extract. Preincubation with the M-peptide completely reversed the blocking effect, indicating that the elimination of the normal retarded band is indeed due to the specific antigen recognition (Figure 7). Identical results were seen with preadipocyte extracts.

We also wished to ask whether the effect of

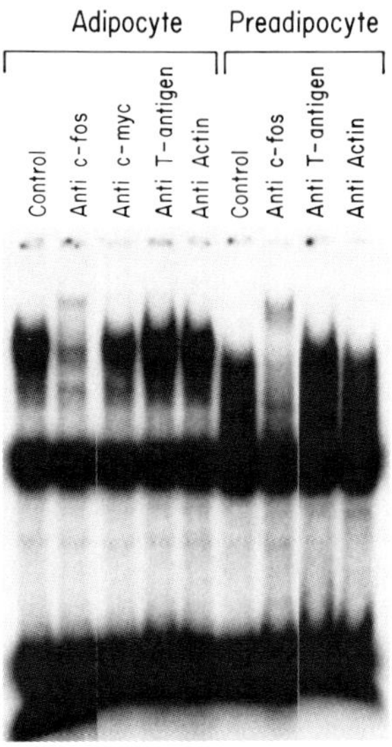

Figure 6. c-fos Antibody Prevents aP2 FSE2 Nucleoprotein Complex Formation. Adipocyte and preadipocyte nuclear extracts (3 µg) were preincubated with no addition (Control) or with the addition of purified antibodies to c-fos (M-peptide, 0.5 µg), c-myc (2.2 µg), SV40 large T antigen (1.2 µg), or actin (0.88 µg). Antibodies are described in Distel et al. (1987). Radiolabeled aP2 FSE2 oligomer was then added, and the complexes were analyzed on 4% nondenaturing gels. N denotes a retarded band that is not sequence-specific for FSE2 DNA.

the anti-c-fos antibodies was general for all DNA-binding complexes. Previous work has demonstrated that the GPD FSE2 and the adipsin FSE2-like sequences are recognized in a sequence-specific fashion by nuclear factors that are not competed by the aP2-FSE2 sequence (Distel and Spiegelman, unpublished data). The same anti-c-fos preparation was preincubated with the adipocyte extracts and DNA-protein complex formation on the GPD FSE2 and the adipsin FSE2-like sequences was

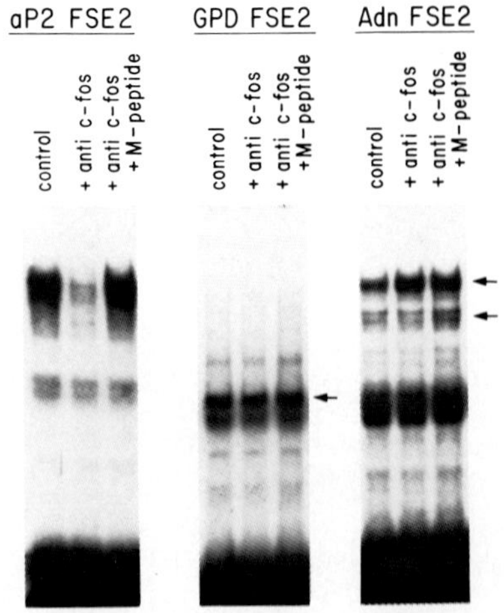

Figure 7. Specificity of the Anti-c-fos Action of FSE2. Adipocyte nuclear extract was preincubated with no addition (control) or with either 0.5 μg of anti-M-peptide antibodies (+ anti-c-fos) or 0.5 μg of anti-c-fos antibodies that had been preincubated with 5 μg of the cognate M-peptide (+ anti-c-fos + M-peptide). Complexes were then formed with synthetic DNA duplexes encoding the aP2 FSE2, the GPD FSE, and the adipsin FSE2 like synthetic oligomer. The complexes were analyzed on 4% nondenaturing gels. Bands from the GPD or adipsin DNA fragments that are a result of sequence-specific binding were determined by a series of competition studies and are denoted with arrows.

examined. The sequence-specific binding to these fragments was not affected by the anti-c-fos antibodies indicating some degree of specificity for the aP2 FSE2 complex (Figure 7 center and right panels).

We also asked whether the complexes could be disrupted by the anti-c-fos antibodies after they had been formed. Complexes were formed and then incubated with the antibodies for 60 min.

Figure 8 demonstrates that these antibodies were able to disrupt the adipocyte aP2 FSE2 DNA protein complexes and generate the slower migrating bands as seen in the preincubation experiments. These additional bands were shown to be sequence-specific by preventing their formation with the addition of excess unlabeled aP2 FSE2 during the formation of the complex. This ability to disrupt preformed complexes was demonstrated both with the anti-c-fos M-peptide antibody and two separate rabbit antisera raised against a fusion protein between bovine growth hormone and the C-terminal half of human c-fos. Both rabbit antisera react strongly with c-fos but have no detectable binding to other c-fos like proteins (Cochran et al., 1984; Curran et al., 1985; D. Slamon, personal communication; T. Curran, personal communication). This strengthens the notion that c-fos itself is part of the aP2 FSE2 complex.

To examine the physical proximity of c-fos to the aP2 FSE2 DNA, we asked whether it could be directly photo-crosslinked to this sequence and subsequently immunoprecipitated. Figure 9, lane A demonstrates that UV cross-linking of nuclear protein complexes to radiolabeled FSE2 DNA generates two major species when analyzed on SDS-polyacrylamide gels: a heterogeneous smear from 42 to 68 kd and a band of about 22 kd. This pattern is derived from the adipocyte complex seen in non-denaturing gels as it can be seen when the cross-linked material is first run on a non-denaturing gel and the bands excised and run on SDS-polyacrylamide gels. Furthermore, both bands on the denaturing gel are sensitive to competition with homologous DNA but not heterologous DNA. To determine whether either of these photo-crosslinked bands contained c-fos, immunoprecipitations were performed with two different antibodies to c-fos. Antibodies to the c-fos fusion protein

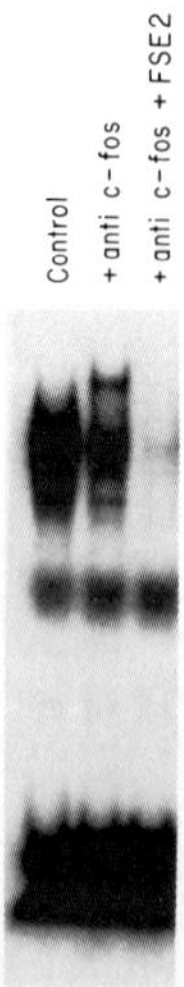

Figure 8. c-fos Antibody Disrupts Preformed FSE2 Nucleoprotein Complexes. aP2 FSE2 DNA-protein complexes were preformed for 30 min, and anti-M-peptide antibodies were added (+ anti-c-fos) and incubated for 60 min before electrophoresis. No additions were made to controls. Complexes were also preformed in the presence of cold aP2 FSE2 (10 ng), and then anti-c-fos antibodies were added (+ anti-c-fos + FSE2). Note that the incubation time with antibodies (1 hr) was less than the apparent half-time of complex dissociation (2 hr), indicating that antibodies had a destabilizing effect on preformed complexes.

were added directly to the radiolabeled cross-linked complex, incubated overnight and precipitated with protein A-Sepharose. The precipitated complex was washed with RIPA buffer containing 0.1% SDS. Figure 9, lane B shows the immunoprecipitation of a broad band at 45 to 50 kd, but no precipitation of the 22 kd material. Figure 9, lane C, control serum (prepared against adipsin peptide) precipitates nothing.

c-fos is found in tight but noncovalent association with a phosphoprotein p39 (Curran et al., 1985). In order to determine whether the

DNA cross-linking was occurring directly to c-fos or was occurring via cross-linking to p39 or a similar protein, the immunoprecipitation was

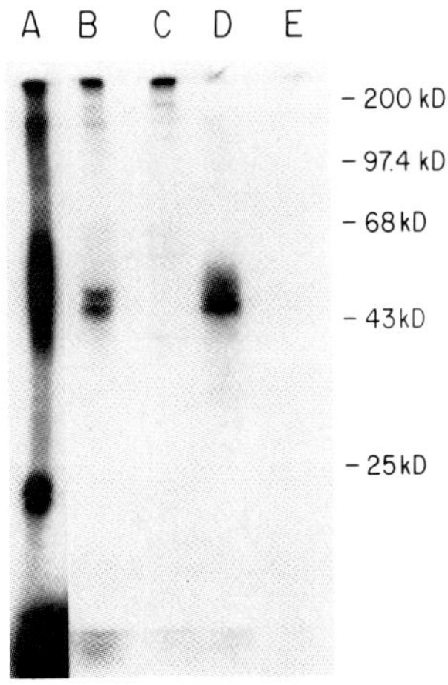

Figure 9. Immunoprecipitation of Photo-Cross-Linked aP2 FSE2 Complexes with Anti-c-fos Antibodies. Photo-cross-linking of ^{32}P-labeled FSE2 complexes and immunoprecipitations were performed as described in Distel et al. (1987). These were then electrophoresed on 10% SDS-polyacrylamide gels (Laemmli, 1970) and autoradiographed. (A) Photo-cross-linked complex before immunoprecipitation. (B) Same as (A), immunoprecipitated with rabbit antisera to c-fos bovine growth hormone fusion protein. (C) Same as (A), immunoprecipitated with control rabbit antisera to mouse adipsin. (D) Same as (A), but boiled in 1% SDS and 5% β-mercaptoethanol for 5 min before immunoprecipitation with anti-c-fos (M-peptide) antibodies. (E) same as (D), but immunoprecipitated with antibodies to SV40 large T antigen. Protein molecular weight markers were BRL high molecular weight range stained with Coomassie (brilliant) blue.

conducted after the cross-linked complex was boiled in SDS and beta-mercaptoethanol, conditions that have been shown to dissociate the two proteins (Curran et al., 1985). The anti-m-peptide antibody was used (Figure 9, lane D) along with the control antibody to SV40 large T-antigen (lane E). The anti-c-fos M-peptide antibody precipitated a complex that ran from 45 to 55 kd. The control antibodies brought down no radiolabeled DNA. These results demonstrate the specificity of the interaction between c-fos and aP2 FSE2 DNA in two ways: only the 45 and 55 kd material was precipitated, and the precipitation was observed only with the antibodies to c-fos. The size of the immunoprecipitated cross-linked product is generally consistent with the size of immunoprecipitated c-fos observed by SDS-PAGE (55 kd; Curran et al., 1984).

DISCUSSION

The FSE2-binding Protein

Previous work from our laboratory has strongly suggested that the FSE2 element plays an important role in the differentiation-dependent control of the aP2 gene, acting as a negative element in preadipocytes but not in adipocytes (Distel et al., 1987). The results presented here and in Distel et al. (1987) indicate that aP2 FSE2 represents a binding site for a sequence-specific nuclear protein in both preadipocytes and adipocytes. The patterns of protein binding change drastically by gel retardation assay during differentiation from a broad smear in preadipocytes to a sharp band in adipocytes. Several lines of evidence suggest that the smear seen in preadipocytes is not due to dissociation of the complexes but reflects a heterogeneity of nucleprotein complexes. The preadipocyte complexes have a comparable half-life to the adipocyte complexes in solution. Experiments not shown indicate that the adipocyte and preadipocyte complexes are equally stable in the polyacrylamide gel. Rerunning parts of the preadipocyte smear after excision shows that each part of the smear is made up of nucleoprotein complex, not free DNA as would be expected if the complexes were dissociating in the gel. Since mobility in these gels appears to be primarily a function of protein mobility, the preadipocyte smear probably reflects the heterogeneous nature of the protein moiety. This heterogeneity obviously could reflect changes in the number or nature of the protein subunits bound to the DNA, including very extensive protein modifications, covalent or noncovalent. While changes in protein conformation cannot be ruled out entirely, they would have to be so stable relative to the time of electrophoresis (2 hours) that this possibility appears remote. It remains to be seen whether the changes observed in the protein binding pattern are causally related to the altered activity of the aP2 FSE2 site during differentiation.

Participation of c-fos in the FSE2 Nucleoprotein Complex.

The evidence presented here indicates that c-fos is an important element of the aP2 nucleoprotein complex in both adipocytes and preadipocytes. Three separate preparations of anti-fos antibodies, one prepared against a peptide of c-fos protein and two against a bacterial fusion protein made with a large segment of c-fos, have the ability to block nucleoprotein complex formation and disrupt complexes already formed. Antisera or purified antibodies to c-myc, SV40 large T antigen, actin or adipsin do not have this property. The effect is due to antigen recognition by the antibodies because the peptide used for immunization (M-peptide) could completely block the disruptive effect. We have demonstrated that this effect is gene-specific by showing that nucleoprotein complexes at the FSE2 of GPD or the FSE2-like site in adipsin appear unaffected by the anti-c-fos antibodies. Finally, two different preparations of antibodies to c-fos could specifically immunoprecipitate radiolabeled DNA covalently linked to protein complexes. Control antisera had no effect.

It is potentially important to understand whether the protein of interest on the aP2 FSE2 site is actually c-fos or another immunochemically similar protein, as proteins with sequence similarity to c-fos have been reported (Cochran et al., 1984) and the anti-peptide antibody used in these experiments has been shown to recognize c-fos-like proteins (Curran et al., 1985; Sambucceti and Curran, 1986). However, the antisera prepared to the c-fos bacterial fusion protein are directed toward a different portion of the protein (C-terminal) and recognize only authentic c-fos in immunoprecipitations from a number of cell lines (T. Curran, personal communication; D. Slamon, personal communication). Thus, the ability of these antisera to react with the FSE2-binding protein suggests that c-fos itself is likely to be a part of these trans-acting complexes. A definitive identification and characterization of this protein, however, must await purification of the FSE2-binding factors.

Do these results give us any insight into how c-fos contributes to the FSE2 nucleoprotein complex? The ability of anti-fos antibodies to disrupt preformed complexes suggests that normal c-fos function is required for these proteins to bind to DNA. When these complexes are disrupted, most of the DNA is released and appears to run with the free oligonucleotide (Figures 6-8). No new bands of higher mobility are observed, and althought a new, slower migrating form is observed, it accounts for no more than 5% of the released DNA. Were c-fos present but not required for binding activity, it would be predicted that antibody binding should shift the majority of counts into a band of higher or lower mobility, depending on whether the antigen dissociated from the rest of the nucleoprotein complex or was retained. This suggests that c-fos is intimately involved in the sequence recognition and DNA-binding properties of the aP2 FSE2 complexes. c-fos alone may have this property, or the complex of c-fos with another protein may represent the DNA binding unit.

No prior evidence has been presented for the direct involvement of c-fos (or any other nuclear proto-oncogene) in the direct control of cellular genes through trans-acting factors, although much indirect evidence points to proto-oncogenes as key regulators of gene activity, especially in cell growth and differentiation. Specifically, c-fos is rapidly and transiently induced in many cell lines (e.g. Greenberg et al., 1985), including 3T3-L1 preadipocytes and adipocytes (Stumpo and Blackshear, 1986), in response to polypeptide growth factors and other growth-inducing factors. In keeping with our observations, others have shown c-fos to be in close association with chromatin in vivo and in vitro and can bind directly to DNA-cellulose (Sambucetti and Curran, 1986; Renz et al., 1987).

Further investigations will be required to explain the wide range of processes associated with c-fos expression and to determine whether c-fos participates as a trans-acting factor at control sites in other genes related to cell growth and differentiation. In any event, it appears that the nucleoprotein complex from the aP2 gene will be useful for detailed analysis of c-fos function.

ACKNOWLEDGEMENTS

We thank the following colleagues for many helpful discussions: Hyo Sung Ro, Barry Rosen, Douglas Groves, Tom Roberts, Debby Dobson and Hye Yeong Min. We also thank Adah Levens for expert preparation of this manuscript. This work has been supported by grants from the National Institute of Health (AM 31405).

Robert Distel is a recipient of an N.R.S.A. Post-Doctoral Fellowship (F32AM07715). Bruce Spiegelman is a recipient of an Established Investigator Award of the American Heart Association.

REFERENCES

Bernlohr DA, Angus CW, Bolanowski MA, Kelly TJ, Jr (1984). Expression of specific mRNA during adipose differentiation: identification of an mRNA encoding a homologue of myelin P2 protein. Proc Natl Acad Sci USA 81:5468-5472.

Bernlohr DA, Bolanowski MA, Kelly TJ, Jr, Lane MD (1985). Evidence for an increase in transcription of specific mRNAs during differentiation of 3T3-L1 adipocytes. J Biol Chem 260:5563-5567.

Chapman AB, Knight DM, Ringold GM (1984). Analysis of gene expression during differentiation of adipogenic cells in culture: hormonal control of the developmental program. J Biol Chem 259:15548-15555.

Cochran BH, Zullo J, Verma IM, Stiles CD (1984). Expression of the c-fos gene and of a fos-related gene is stimulated by platelet derived growth factor. Science 226:1080-1082.

Cook KS, Groves DL, Min HY, Spiegelman BM (1985a). A developmentally regulated mRNA from 3T3 adipocytes encodes a novel serine protease homologue. Proc Natl Acad Sci USA 82:6480-6484.

Cook KS, Hunt CR, Spiegelman BM (1985b). Developmentally regulated mRNA in 3T3-adipocytes: analysis of transcriptional control. J Cell Biol 100:514-520.

Cook KS, Min HY, Johnson D, Chaplinsky R, Flier JS, Hunt C, Spiegelman BM (1987). Adipsin: a novel, circulating serine protease homologue secreted by adipose tissue and sciatic nerve. Science 237:402-405.

Curran T, Miller AD, Zokas L, Verma IM (1984). Viral and cellular fos proteins: a comparative analysis. Cell 36:259-268.

Curran T, Van Beveren C, Ling N, Verma IM

(1985). Viral and cellular fos proteins are complexed with a 39,000-dalton cellular protein. Mol Cell Biol 5:167-172.

Dignam JD, Lebowitz RM, Roeder RG (1983). Accurate transcription initiation by RNA polymerase II in soluble extract from isolated mammalian nuclei. Nucl Acids Res 11:1475-1489.

Distel RJ, Ro HS, Rosen BS, Groves DL, Spiegelman BM (1987). Nucleoprotein complexes that regulate gene expression in adipocyte differentiation: direct participation of c-fos. Cell 49:835-844.

Djian P, Phillips M, Green H (1985). The activation of specific gene transcription in adipose conversion of 3T3 cells. J Cell Physiol 124:554-556.

Doglio A, Dani C, Grimaldi P, Ailhaud G (1986). Growth hormone regulation of the expression of differentiation-dependent genes in preadipocyte Ob1771 cells. Biochem J 238:123-129.

Fried M, Crothers DM (1981). Equilibria and kinetics of Lac repressor operator interactions by polyacrylamide gel electrophoresis. Nucl Acids Res 9:6506-6525.

Garner MM, Revzin A (1981). A gel electrophoresis method for quantifying the binding of proteins to specific DNA regions: application to components of the E. coli lactose operon regulatory system. Nucl Acids Res 9:3047-3060.

Goodbourn S, Burstein H, Maniatis T (1986). The human β-interferon gene enhancer is under negative control. Cell 45:601-610.

Green H (1979). Adipose conversion: a program of differentiation. In Ailhaud G, (ed): "Obesity, Cellular and Molecular Aspects," Paris: Editions INSERM, pp 15-25.

Greenberg ME, Ziff EM (1984). Stimulation of 3T3 cells induces transcription of the c-fos protooncogene. Nature 311:433-438.

Greenberg ME, Green LA, Ziff EM (1985). Nerve growth factor and epidermal growth factor induce rapid transient changes in proto-oncogene transcription in PC12 cells. J Biol

Chem 260:14101-14110.
Hunt CR, Ro JHS, Dobson D, Min HY, Spiegelman BM (1986). Adipocyte P2: developmental expression and homology of 5'-flanking sequences among fat cell-specific genes. Proc Natl Acad Sci USA 83:3786-3790.
Kruijer W, Cooper JA, Hunter T, Verma IM (1984). PDGF induces rapid but transient expression of the c-fos gene. Nature 312:711-716.
Laemmli UK (1970). Cleavage of structural proteins during the assembly of the head of bacteriophage T4. Nature 227:680-685.
Maniatis T, Fritsch EF, Sambrook J (1982). Molecular Cloning: A Laboratory Manual (Cold Spring Harbor, New York: Cold Spring Harbor Laboratory), pp 113.
Min HY, Spiegelman BM (1986). Adipsin, the adipocyte serine protease: gene structure and control of expression by tumor necrosis factor. Nucl Acids Res 14:8879-8892.
Morgan JI, Curran T (1986). The role of ion flux in the control of c-fos. Nature 322:552-555.
Muller R, Bravo R, Burckhardt J, Curran T (1984). Induction of c-fos gene and protein by growth factors precedes activation of c-myc. Nature 312:716-720.
Phillips M, Djian P, Green H (1986). The nucleotide sequence of three genes participating in the adipose differentiation of 3T3 cells. J Biol Chem 261:10821-10827.
Renz M, Verrier B, Kurz C, Muller R (1987). Chromatin associated and DNA binding properties of the c-fos proto-oncogene product. Nucl Acids Res 15:277-292.
Rosenthal N, Kress M, Khoury G (1983). BK-virus enhancer element and human cellular homologue. Science 222:749-755.
Sambucetti L, Curran T (1986). The fos protein complex is associated with DNA in isolated nuclei and binds to DNA cellulose. Science 234:1417-1419.
Sidhu RH (1979). Two-dimensional electrophoretic analyses of proteins synthesized during differentiation of 3T3 L1 preadipocytes. J Biol Chem 254:11111-11118.

Spiegelman BM, Green H (1980). Control of specific protein biosynthesis during the adipose conversion of 3T3 cells. J Biol Chem 255:8811-8818.
Spiegelman BM, Frank M, Green H (1983). Molecular cloning of mRNA from 3T3 adipocytes. J Biol Chem 258:10083-10089.
Stumpo D, Blackshear BJ (1986). Insulin and growth factor effects on c-fos expression in normal protein kinase C-deficient 3T3-L1 fibroblasts and adipocytes. Proc Natl Acad Sci USA 83:9453-9457.
Weber F, de Villiers J, Schaffner W (1984). An SV40 "enhancer trap" incorporates exogenous enhancers or generates enhancers from its own sequences. Cell 36:983-992.
Zezulak KM, Green H (1985). Specificity of gene expression in adipocytes. Mol Cell Biol 5: 419-421.

Cellular Factors in Development and Differentiation:
Embryos, Teratocarcinomas, and Differentiated Tissues, pages 211–224

CHARACTERIZATION OF THE *CIS* AND *TRANS* ELEMENTS ESSENTIAL FOR RAT INSULIN II GENE EXPRESSION

David T. Crowe, Young-Ping Hwung, Sophia Y. Tsai and Ming-Jer Tsai
Department of Cell Biology, Baylor College of Medicine, One Baylor Plaza, Houston, TX 77030

ABSTRACT

We have constructed 14 different linker-scanner (LS) mutants throughout the enhancer and promoter of a rat insulin II-CAT fusion gene. These LS mutants were transiently transfected into an insulin-producing (HIT) cell line and three mutants (LS-261/252, LS-102/91, LS-54/45) displayed drastically reduced levels of CAT activity. Therefore, at least three regions are essential for the *in vivo* expression of an insulin-CAT fusion gene.

To identify the *trans*-acting factors which interact with this cell-specific promoter, we performed band-shifting assays with either HIT or HeLa nuclear extracts and an end-labelled insulin fragment (-100 to +49). Three binding activities were common to both extracts, and another one was unique to HIT cells. DNase I protection studies localized one binding activity to -60 to -40 bp 5' to the cap site. We show that this factor is common to both cell lines and is identical to a previously characterized transcription factor (COUP) which binds to the chicken ovalbumin upstream promoter.

INTRODUCTION

A common goal in molecular biology is to understand the mechanisms which govern the temporal and tissue-specific transcription of genes. Recent efforts have used mutagenesis to disrupt target genes and their reintroduction into cells have identified a class of regulatory elements which function when coupled in *cis* 5', 3', or within actively transcribed genes. While the physical interactions of *trans*-acting factors with enhancers and promoters of "housekeeping" and regulated genes have been described, the mechanisms by which these complexes initiate and regulate gene expression are not well understood. Therefore, in order to understand how genes are both positively and negatively regulated during development and differentiation, a more comprehensive analysis of *cis*-acting elements is needed along with the use of purified transcription factors to study protein-DNA interactions and in vitro transcription.

We have selected the rat insulin II gene as a model system for tissue-specific gene expression. Insulin gene expression is restricted to the B cells of the islets of Langerhans in the pancreas. Insulin is essential for glucose homeostasis and is precisely regulated primarily in response to various nutrients in the bloodstream. The rat genome contains two nonallelic insulin genes (I and II) (Lomedico et al., 1979; Cordell et al., 1979) and quantitation of the biosynthesis of insulin indicates that both genes are equally expressed (Clark & Steiner, 1969).

The *cis*-linked sequences essential for proper cell-specific expression of the rat insulin gene have been defined by the introduction of insulin-reporter fusion genes into either the mouse genome or into insulin-producing cell lines. Transgenic mice bearing 520 bp of a rat insulin II promoter fused to the coding sequences of the SV40 large T antigen developed only B cell tumors; since this 5'-flanking region displayed targeted expression of the oncogene regardless of its orientation, it suggests that a cell-specific enhancer is present (Hanahan, 1985). Furthermore, the transient transfection of a rat insulin I-CAT (chloramphenicol acetyltransferase) fusion gene into a hamster insulinoma (HIT) cell line indicated that sequences within the 5'-flanking region were sufficient to direct proper cell-specific expression (Walker et al., 1983). Additionally, a cell-specific enhancer (-333 to -103) and promoter (-113 to +1) have been identified by promoter and enhancer replacement studies (Edlund et al., 1985). Since the two rat insulin genes are very similar (~85%) (Fig. 3; Soares et al., 1985) throughout the 400 bp 5' to the transcription initiation site, it is very likely that the corresponding sequences in the rat insulin II gene also have enhancer and promoter activity.

Silencers are *cis*-linked elements which have many of the features of enhancers except that they serve to diminish transcription rather than augment it. There appear to be two such negative regulatory elements in the rat insulin I gene. One is located between -4.0 to -2.0 kb upstream of the cap site and is a member of a long interspersed rat repetitive element (Laimmons et al., 1986). A second element occurs downstream of -410 (Nir et al., 1986). It is possible that these elements function to repress insulin expression in inappropriate cell types, while in insulin producing cells, some interplay between positive and negative elements might be required to facilitate proper cell-specific expression.

Since many enhancers and promoters appear to consist of multiple arrays of short sequence motifs (Zenke et al., 1986; Schirm et al., 1987), one of our goals is to identify the 5'-flanking sequences in the rat insulin II gene that are essential for its expression. Preliminary results indicate that there are at least three regions which are essential for the expression of the rat insulin II gene. We also seek to characterize those factors which interact with either the enhancer or promoter, and are currently characterizing a transcription factor common to HIT and HeLa cells which binds to the promoters of both the rat insulin II and chicken ovalbumin genes.

METHODS AND MATERIALS

We constructed linker-scanning (LS) mutants by recombining a series of 3'- and 5'-deletions (Haltiner et al., 1985) occurring throughout the 5'-flanking region of the rat insulin II gene. All LS mutants contained the sequences between -448 to +49 (Eco RI to Rsa I fragment) fused to the CAT gene-SV40 splice and poly (A) addition sites found in pSV2CAT (Gorman et al., 1982). Fourteen LS mutants were made throughout the enhancer and promoter in which an 8 bp Sac I linker replaces 5-11 bps of wild-type sequence (Fig. 1). Double-stranded sequencing with a ^{32}P-end-labelled sequencing primer (McGraw, 1984) was use to determine the precise endpoints of all 5'- and 3'-

deletions while restriction mapping was used to confirm the location of the newly introduced Sac I site.

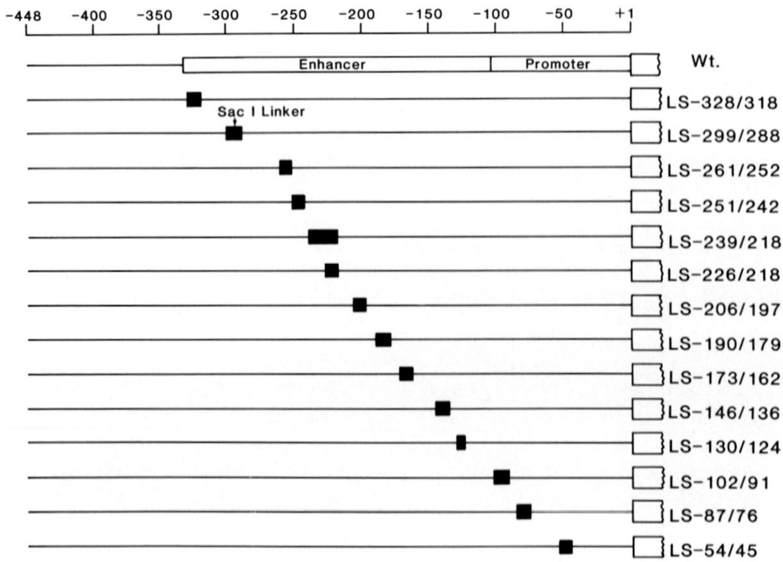

Figure 1. The location of the Sac I linker in each LS mutant is depicted by a black box. There are 14 different linker-scanners with mutations throughout the enhancer and promoter of the rat insulin II-CAT gene.

We performed transient transfection experiments according to existing protocols (Edlund et al., 1985) except that 5 µg of test DNA were applied to a 6 cm dish containing a hamster insulinoma (HIT) cell line (Santerre et al., 1981). Although cells were routinely maintained in RPMI 1640, 10% fetal calf serum, and 1% (v/v) of a penicillin/streptomycin solution (GIBCO Laboratories, Grand Island, N.Y.), this medium was replaced with DMEM minus serum during the time of DNA addition until the glycerol shock. Assays for CAT activity were performed 48 to 60 hours post-transfection (Gorman et al., 1982).

The preparation of cell extracts, gel retardation assays, DNase I protection studies, and the purification of the COUP transcription factor were performed as previously described (Sagami *et al*., 1986). The techniques used in the DNA-affinity chromatography and renaturation of the COUP transcription factor have also been described (Wang *et al*., 1987).

RESULTS AND DISCUSSION

Three Regions are Essential for Rat Insulin II Expression

We transfected fourteen LS mutants into HIT cells and identified three different regions essential for the expression of a rat insulin II-CAT hybrid gene (Fig. 2).

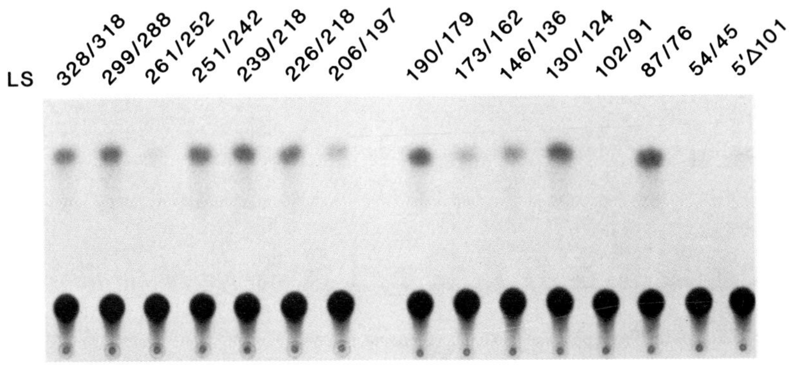

Fig. 2. LS mutants were transiently transfected into HIT cells and cell extracts were assayed for CAT activity. LS-261/252, LS-102/91, and LS-54/45 reduce CAT activity to levels comparable to an enhancerless insulin-CAT construct (i.e. a 5'-deletion to -101).

These mutants (LS-261/252, LS-102/91, and LS-54/45) have an 8 bp Sac I linker which replaces the natural sequences between -261 to -252, -102 to -91, and -54 to -45, respectively. In these experiments, since the body of the mRNA isn't altered in the mutant constructs, the level of CAT activity should reflect the influence of enhancer or promoter mutations on the rate of transcription. These three mutants reduced CAT activity to essentially background levels equivalent to that obtained by transfecting with a promoterless CAT construct (pSVOCAT) or a plasmid containing a 5'-deletion in the rat insulin II gene to -101. One of these mutants, LS-261/252, falls within the enhancer, while LS-102/91 lies near the junction of the enhancer and promoter, and LS-54/45 is found in the promoter region.

Sequence comparison of the rat insulin II 5'-flanking region with other viral and eucaryotic enhancers and promoters reveals numerous motifs (especially if allowances are made for slight mismatches to the consensus), however their functional role in this context is unknown. For example, LS-102/91 disrupts a region which is very similar to the E motif (CAGGTGCC vs. CAGATGCC) (Fig. 3) which is found in the enhancer of the Ig heavy chain and kappa light chain (Ephrussi *et al.*, 1985). Generally, even though several other LS mutants appear to disrupt potential motifs, most do not greatly perturb transcriptional activity. Another example is found in the rat insulin I gene which contains a near perfect consensus to the SV40 core enhancer at -316 to -305 which can be removed by a 5'-deletion without a drastic loss of enhancer activity (Edlund *et al.*, 1985). One possible explanation of these results is that a high level of functional redundancy exists among these motifs, such that many can be individually removed without significant effect due to compensatory action by the others. This appears to be the case for the Ig heavy chain enhancer, where mutations in any one E or octanucleotide motif reduce activity by not more than two-fold, while combinations of E mutations had deleterious results (Lenardo *et al.*, 1987). This does not preclude the possibility that some elements are more essential than others in regulating correct expression of the gene.

Clearly, the sequences between -261 to -252, -102 to -91, and -54 to -45 play a critical role in the expression of the rat insulin II gene since mutations in these regions reduce CAT activity to levels comparable to the transfection of an insulin gene without its enhancer. Except for those sequences between -102 to -91, the sequences throughout the two other regions are not highly conserved even among other mammalian insulin genes

(Soares et al., 1985), although these regions do appear in the enhancers and promoters of a number of other viral and eucaryotic genes. We do not know which (if any) of these or other individual motifs endows cell-specificity

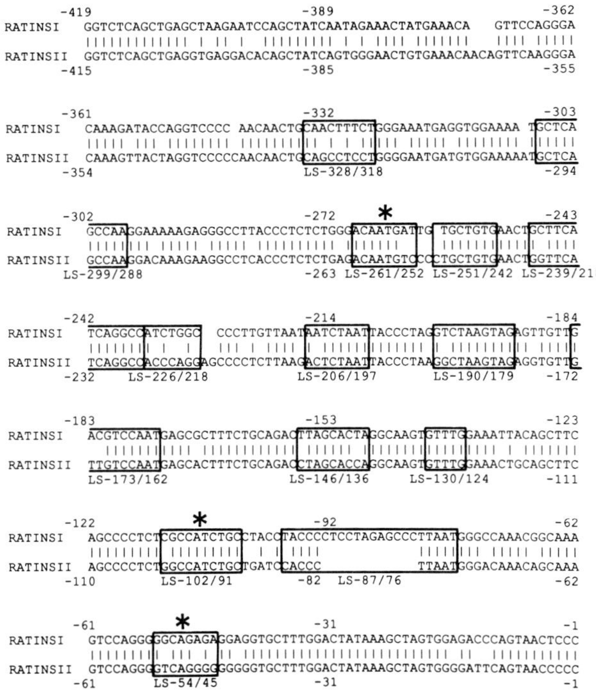

Figure 3. Sequence alignment of the 5'-flanking region of the rat insulin I and II genes. The sequences of the rat insulin II gene within the boxes were replaced with an 8 bp Sac I linker (CGAGCTCG). Starred boxes denote those LS mutants which displayed greatly reduced CAT activity.

in the enhancers and promoters of a number of other viral and eucaryotic genes. We do not know which (if any) of these or other individual motifs endows cell-specificity upon the insulin enhancer and promoter or if its specificity results from a combination of several generally occurring *cis*-linked elements. To address this question, we plan to couple individual or concatemerized motifs to a "neutral" promoter and then test if these elements confer upon the reporter gene the preferential ability to be expressed in HIT cells.

A Transcription Factor Present in Both HIT & HeLa Extracts Binds to the Rat Insulin II Promoter

Having defined three regions which were essential for rat insulin II gene expression, we then sought to identify and characterize those *trans*-acting factors which interacted with these elements. We initially examined those sequences around -54 to -45 which reside within the cell-specific insulin promoter. Both HIT and HeLa nuclear extracts were tested for the presence of factors which bound to this region by using a band-shifting assay. A ^{32}P-end-labelled probe which spanned the promoter (-100 to +49) was incubated with nuclear extracts from either cell type and the protein-DNA complexes were resolved on native polyacrylamide gels. We identified three shifted bands (#1, 2, and 4) which were common to both extracts while an additional band (#3) appears to be restricted to HIT cells (Fig. 4).

When HeLa extracts were utilized in DNase I-footprinting studies to localize the site of protein-DNA interactions within the promoter region, a region from -60 to -40 bp upstream of the cap site was protected from digestion (Fig. 5). An identical region was also protected when HIT extracts were used (Hwung et al., manuscript in preparation). It is now clear that a factor is present in both HIT and HeLa cells which binds to sequences within the rat insulin II promoter in a region which is essential for the expression of this gene.

During further fractionation of HeLa cell extracts, we found that this footprinting activity co-purified with a previously characterized transcription factor, COUP, that is also found in HeLa cells, and binds to the chicken ovalbumin upstream promoter at -90 to -70 (Sagami et al., 1986). The COUP binding protein is required for the efficient *in vitro* transcription of the ovalbumin gene and binds to the sequence 5'-GGTGTCAAAGGTCAAACT-3'.

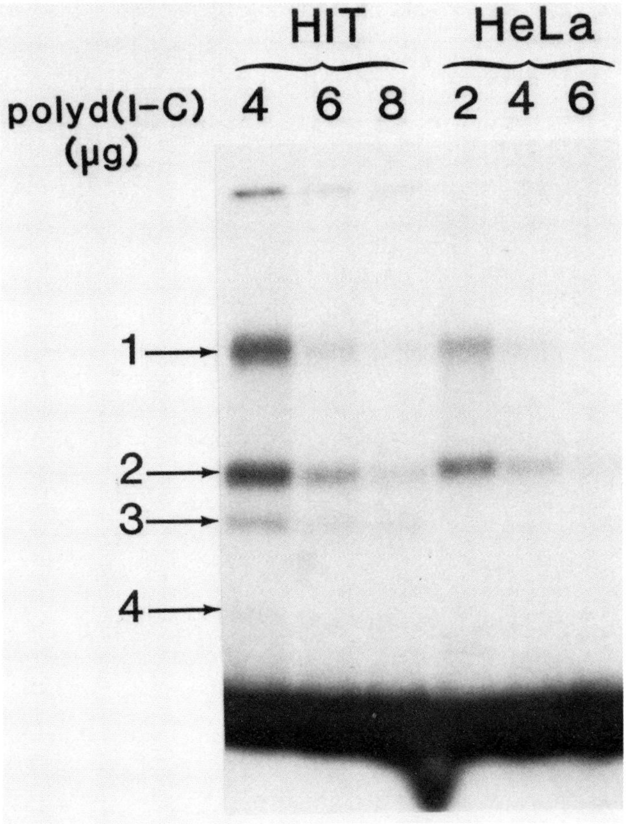

Figure 4. Band-shifting assays were performed using an end-labelled rat insulin II fragment (-100 to +49) with HIT or HeLa extracts. Binding activities to the promoter-bearing fragment are present in both extracts (see shifted bands #1, 2, and 4), whereas the binding activity associated with band #3 appears to be restricted to HIT cells.

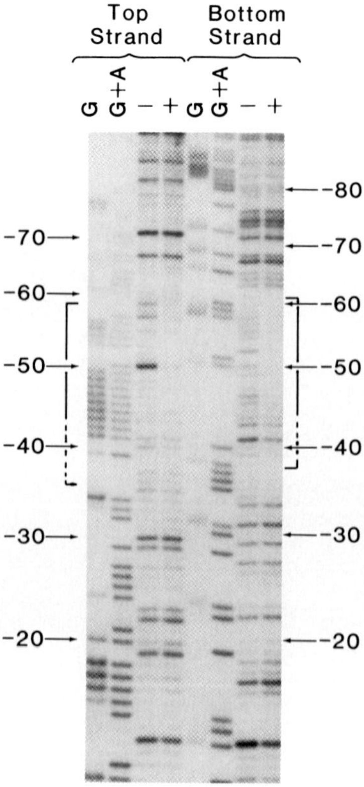

Figure 5. HeLa cell extracts were used in DNase I-protection studies of the rat insulin II promoter. "-" and "+" refers to the absence or presence of extract.

DNA-affinity chromatography was then used to purify the COUP transcription factor to near homogeneity, and electrophoresis of this material on an SDS-polyacrylamide gel revealed the existence of 5 major bands after silver staining (Fig. 6). The gel was then cut into 5 mm slices and the proteins contained in these slices were electro-eluted and renatured and used in band-shifting assays with either an ovalbumin or insulin promoter-bearing fragment. Proteins renatured from slices 3 to 6 contained polypeptides spanning from 43 to 53 kD, which were able to bind specifically to both insulin and ovalbumin promoter sequences. These experiments along with binding competition assays (Hwung et al., manuscript in preparation) suggest that the COUP transcription factor is capable of binding to the promoters of two distinctly different tissue-specifically expressed genes.

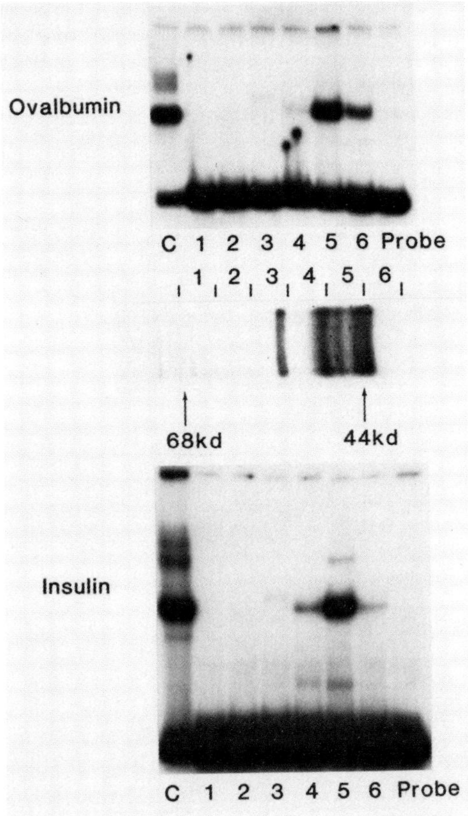

Figure 6. An SDS-polyacrylamide gel containing DNA-affinity purified COUP was cut into six-5 mm slices. Polypeptides were electroeluted, renatured, and used in gel retardation assays. Polypeptides in slices 3, 4, 5, and 6 contained band-shifting activity.

Several factors may be interacting with the rat insulin II promoter as evidenced by our band-shifting experiments, and we are currently defining the specific sites at which these additional factors bind to the promoter. Additional band-shifting experiments using purified COUP suggest that of the four shifted bands observed with nuclear extracts, band #2 represents the binding of this transcription factor to the promoter. Although this promoter is cell-specific in its expression, it is unlikely that the COUP binding protein is mediating this specificity by itself, since the presence of COUP is not restricted solely to B cells. A more likely possibility is that COUP functions in a combinatorial manner with a number of other factors to regulate gene expression in a cell-specific manner.

It therefore appears that the rat insulin II gene along with the chicken ovalbumin, ovomucoid, and Y genes belongs to a class of genes which require the COUP box binding protein for their efficient expression, while the SV40 early, adenosine deaminase, thymidine kinase, lysozyme, MMTV, and β-globin genes do not appear to have such a requirement (Sagami _et al_., 1986; Tsai _et al_., 1987). This transcription factor is distinct from CTF (Jones _et al_., 1987), which is also found in HeLa cells and binds to the CAAT box of several eucaryotic and viral promoters, although the COUP and CAAT box binding proteins could share a common mechanism in transcriptional activation.

ACKNOWLEDGEMENTS

The authors wish to thank the following for gifts and assistance in this project: Dr. A.E. Boyd (HIT cells), Dr. R. Tjian (plasmid pBRN/S), Dr. P. Lomedico (rat insulin II genomic clones), Dr. L-H Wang (purification and renaturation of COUP), and Mr. C. Wilkinson (plasmid preparation). This work is supported by grants from the N.I.H. (#HD-17379) and the Baylor Diabetes Center.

REFERENCES

Clark JL, Steiner DF (1969). Insulin biosynthesis in the rat: demonstration of two proinsulins. Proc Natl Acad Sci USA 62:278.

Cordell B, Bell G, Tischer E, DeNoto FM, Ullrich A, Pictet R, Rutter WJ, Goodman HM (1979). Isolation and characteriztaion of a cloned rat insulin gene. Cell 18:533.

Edlund T, Walker MD, Barr PJ, Rutter WJ (1985). Cell-specific expression of the rat insulin gene: evidence for role of two distinct 5'-flanking elements. Science 230:912.

Ephrussi A, Church GM, Tonegawa S, Gilbert W (1985). B lineage-specific interactions of an immunoglobulin enhancer wtih cellular factors in vivo. Science 227:134.

Gorman CM, Moffat LF, Howard BM (1982). Recombinant genomes which express chloramphenicol acetyltransferase in mammalian cells. Mol Cell Biol 2:1044.

Haltiner M, Kempe T, Tjian R (1985). A novel strategy for constructing clustered point mutations. Nucl Acids Res 13:1015.

Hanahan D (1985). Heritable formation of pancreatic β-cell tumors in transgenic mice expressing recombinant insulin/simian virus 40 oncogenes. Nature 315:115.

Laimins L, Holmgren-Konig M, Khoury G (1986). Transcriptional "silencer" element in rat repetitive sequences associated with the rat insulin 1 gene locus. Proc Natl Acad Sci USA 83:3151.

Lenardo M, Pierce JW, Baltimore D (1987). Protein-binding in Ig gene enhancers determine transcriptional activity and inducibility. Science 236:1573.

Lomedico P, Rosenthal N, Efstradiatis A, Gilbert W, Kolodner R, Tizard R (1979). The structure and evolution of the two nonallelic rat preproinsulin genes. Cell 18:545.

McGraw RA (1984). Dideoxy DNA sequencing with end-labeled oligonucleotide primers. Anal Biochem 143:298.

Nir U, Walker MD, Rutter WJ (1986). Regulation of rat insulin 1 gene expression: evidence for negative regulation in nonpancreatic cells. Proc Natl Acad Sci USA 83:3180.

Sagami I, Tsai SY, Wang H, Tsai M-J, O'Malley BW (1986). Identification of two factors required for transcription of the ovalbumin gene. Mol Cell Biol 5:4259.

Santerre RF, Cook RA, Crisel RMD, Sharp JD, Schmidt RJ, Williams DC, Wilson CP (1981). Insulin synthesis in a clonal cell line of simian virus 40-transformed hamster pancreatic beta cells. Proc Natl Acad Sci USA 78:4339.

Schirm S, Jiricny J, Schaffner W (1987). The SV40 enhancer can be dissected into multiple segments, each with a different cell type specificity. Genes and Devel 1:65.

Soares MB, Schon E, Henderson A, Karathanasis SK, Cate R, Zeitlin S, Chirgwin J, Efstratiadis A (1985). RNA-mediated gene duplication: the rat preproinsulin I gene is a functional retroposon. Mol Cell Biol 5:2090.

Tsai, SY, Sagami I, Wang H, Tsai M-J, O'Malley BW (1987) Interactions between a DNA-binding transcription factor (COUP) and a non-DNA binding factor (S-300-II). Cell 50:701.

Walker MD, Edlund T, Boulet AM, Rutter WJ (1983). Cell-specific expression controlled by the 5'-flanking region of insulin and chymotrypsin genes. Nature 306:557.

Wang L-H, Tsai SY, Sagani I, Tsai M-J, O'Malley BW (1987). Purification and characterization of COUP transcription factor from HeLa cells. JBC (in press).

Zenke M, Grundstrom T, Matthes H, Wintzerith M, Schatz C, Wilderman A, Chambon P (1986). Multiple sequence motifs are involved in SV40 enhancer function. EMBO J 5:387.

Cellular Factors in Development and Differentiation:
Embryos, Teratocarcinomas, and Differentiated Tissues, pages 225–241

TRANSCRIPTIONAL REGULATION OF COMPETENCE GENES IN BALB/C-3T3 CELLS

Brent H. Cochran, John Townes, Timothy E. Hayes

Center for Cancer Research and Department of Biology, Massachusetts Institute of Technology, Cambridge, MA 02139

INTRODUCTION

The growth of mammalian cells is strictly regulated by a variety of extracellular growth regulatory factors. These growth factors can be broadly grouped into three categories (Cochran, 1985). One group are called competence factors and are exemplified by the platelet-derived growth factor (PDGF) and the fibroblast growth factor (FGF) for fibroblasts or concanvalin A and antigen in the case of T lymphocytes. The competence factors are distinguished by the fact that exposure of the appropriate target cells to these factors for a time period short relative to the cell cycle transit time is sufficient for them to exert their mitogenic action (Pledger et al. 1978). They give the signal to cells to prepare to leave the G_0 or quiescent state. However, they are not sufficient or efficient in fully supporting cell proliferation. Efficient cell growth is supported by the combination of competence factors with other growth factors called progression factors. Progression type growth factors can be typified by the insulin-like growth factors IGF-1 for fibroblasts, and by interleukin 2 for T cells (Stiles et al., 1979). Progression factors are required for transit of cells through G_1 into the S-phase of the cell cycle. A third type of growth regulatory factors are in fact growth inhibitors, such as the interferons and transforming growth factor-β. These factors can reversibly antagonize the effects of the growth promoting factors. (Massague, 1987; Clemens and McNurlan, 1985).

A major question in cell biology is how do these polypeptide growth factors function inside the cell to regulate cell growth and proliferation. Results of recent research strongly suggest that the biochemical pathways regulated inside the cell by growth factors serve as targets for the primary genetic lesions in the transformed cell. Thus, some growth factors are directly encoded by retroviral oncogenes. The simian sarcoma virus oncogene v-*sis* encodes a growth factor closely related to the platelet-derived growth factor (PDGF)(Waterfield et al., 1983; Doolittle et al., 1983). Other retroviral oncogenes, v-*erbB*, and v-*fms*, encode mutated versions of the growth factor receptors for EGF and CSF-1 (Downward et al., 1984; Scherr et al., 1985) in such a way that they constitutively activate the protein tyrosine kinase activity of these receptors. Another set of oncogenes, the *ras* oncogenes, appear to be closely related to guanine nucleotide binding proteins, which are thought to couple receptor-ligand complexes to various second-signal generating enzymes at the membrane such as adenylate cyclase (Hurley et al., 1984). Finally, many competence type growth factors such as PDGF can rapidly regulate new gene expression (Cochran et al., 1985; Kelly et al., 1983; Cochran et al., 1984). This set of genes has been termed "competence genes". At least two of the competence genes regulated by PDGF, c-*fos* and c-*myc*, are themselves cellular homologues of retroviral oncogenes.

Previous experiments have strongly implicated the induction of expression of c-*fos* and c-*myc* as being essential to cell-cycle transit. Thus, introduction into Balb/c-3T3 cells of exogenous c-*myc* genes put under the control of a mouse mammary tumor virus promoter now makes the c-*myc* gene inducible by hydrocortisone (Armelin et al, 1984). In such cells, the addition of glucocorticoids now has a growth stimulatory effect. Analagous experiments have been done by directly microinjecting c-*myc* protein into quiescent fibroblasts and, there too, a mitogenic response was seen (Kaczmarek, 1985). In addition, antisense RNA experiments suggest that expression of c-*fos* is necessary for cell-cycle transit (Holt et al., 1986; Nishikura and Murray, 1987). Thus, the elucidation of the molecular pathways by which these two genes are regulated will prove crucial not only to the understanding of growth factor action, but also be fundamental to our understanding of tumorigenesis.

MATERIALS AND METHODS

Cell Culture and Growth Factors

All experiments were performed on Balb/c-3T3 cells (clone A31) which were grown in Dulbecco's modified Eagle medium (DME) supplemented with 10% bovine calf serum. Upon reaching confluence, the medium was changed to DME supplemented with 5% human platelet-poor plasma (Scher et al., 1978). Experiments were intitiated on the following day.

Phorbol 12-myristate 13-acetate (TPA) and cycloheximide were obtained from Calbiochem (San Diego). Basic fibroblast growth factor (b-FGF) was obtained from Collaborative Research (Waltham, MA).

V-*sis* conditioned medium (SCM) was produced from an NRK cell line which harbors a simian sarcoma virus provirus (Thiel et al., 1981; Owen et al., 1984). Conditioned medium was prepared as described in Hayes et al. (1987a) and was used at a concentration which gave a maximum induction of gene expression.

Plasmid DNAs

The c-*fos* plasmid used in the runoff transcription experiments was pc-fos (mouse)-3 (Miller et al., 1984). The mouse c-myc cDNA plasmid was pM c-*myc* 54 (Stanton et al., 1983). The mouse c-*myc* genomic clone was a 3.5 kb BamH1-Hind III fragment of the gene which spans the second and third exon cloned into pBR322. The JB, JE, and KC plasmids were as described in Cochran et al. (1983) except that the cDNA inserts were subcloned into pSP64 plasmids. The vector DNA was pSP64.

Runoff Transcription Assays

Nuclear runoff transcription assays were performed by using a modification of the method of Greenberg and Ziff, 1984. After the appropriate treatments, the cell monolayers from two 15 cm tissue culture dishes were washed twice with ice-cold phosphate-buffered saline(PBS). The cells from both dishes were then scraped into 10 ml of ice-cold PBS and pelleted by centrifugation in an IEC clinical centrifuge at 4°C. The cell pellets were re-suspended in 2 ml of lysis buffer (10 mM Tris, pH 7.4, 10 mM NaCl, 3 mM $MgCl_2$, 0.5% NP-40) and incubated on ice for five minutes. The nuclei were pelleted at top speed in a clinical centrifuge for five minutes at 4°C, resuspended in 2 ml of lysis buffer, and pelleted again. The nuclear pellets were resuspended in 100 μl glycerol buffer (40% glycerol, 50 mM Tris, pH 8.3, 5 mM $MgCl_2$, 0.1 mM EDTA). The nuclei were then centrifuged for 30 seconds at 6500g in a microfuge and the nuclear pellets resuspended again in 100 μl of glycerol buffer with gentle vortexing. The nuclei were stored at -70°C until ready for the runoff transcription reaction.

To initiate the runoff transcription reaction, 110 μl of a 2X reaction buffer containing 10 mM Tris, pH 8.0, 5 mM $MgCl_2$, 0.3 M KCl, 1 mM ATP, 1 mM CTP, 1 mM GTP, 2.5 μM UTP, 5 mM dithiothreitol, and 200 μCi of ^{32}P-UTP at 800 Ci/mmol was added to the frozen nuclei and thawed while vortexing. The reaction was incubated for 30 minutes at 30°C, and terminated by the addition of 5 units of RQ1 DNAse (Promega Biotec). The nuclei were incubated for five minutes at 37°C and then the reaction mixture was added to 3.75 ml of guanidine isothiocyanate lysis buffer (5 M guanidine isothiocyanate, 0.1 mM β-mercaptoethanol, 25 mM NaCitrate, pH 7.0). 50 μg of tRNA was added to the lysis buffer as a carrier and the entire solution was layered onto 1.25 ml of 5.7 M CsCl, 25 mM NaAcetate, pH 7, in a polyallomer tube for an SW50.1 type rotor. The gradient was centrifuged in a Beckman SW50.1 rotor for 18 hours at 40,000 rpm. After the spin, the top of the gradient was aspirated, taking care not to contaminate the pellet with the upper layer. The bottom of the tube was cut away from the top and the RNA pellet was resuspended in a total volume of 300 μl of H_2O and precipitated by the addition of 30 μl of 3 M NaAcetate, pH 5.5 and 1 ml of 95% ethanol at -20°C. The RNAs were pelleted by a ten minute centrifugation in a microfuge, and the resulting pellet was resuspended in 500 μl of low salt hybridization buffer (10 mM TES, pH 7.4, 0.2% SDS, 10 mM EDTA). This solution was boiled for three minutes and two 5 μl aliquots were then taken for TCA precipitation. (Resuspension of the RNA pellets by boiling was found to be necessary for obtaining consistent TCA precipitable counts.) For a given experiment and set of conditions, TCA precipitable counts across the set of samples was normalized such that an equal amount of counts, approximately 5 x 10^6 cpm, would be present in each hybridization reaction. The appropriate number of cpm was then diluted up to 1 ml with low-salt hybridization buffer, boiled for three minutes, and added to 1 ml of a high-salt hybridization buffer (10 mM TES, pH 7.4, 0.2% SDS, 10 mM EDTA, 600 mM NaCl). The resulting hybridization solution was then added to a plastic mini scintillation vial containing a nitrocellulose strip in a spiral configuration to which various plasmid DNAs had been bound.

Plasmid DNAs were bound to the nitrocellulose by the following method: The plasmids were linearized by digestion with appropriate restriction enzymes, extracted with phenol/chloroform (1:1), and ethanol precipitated. The resulting DNA was resuspended in TE (10 mM Tris, pH 8.0, 1 mM EDTA) at a concentration of 2.2 μl per μg of DNA. The DNA was denatured by treatment with 0.1 volumes of 1.0 M NaOH for 30 minutes at room temperature. 100 volumes of 6X SSC was added to this solution and then applied to a nitrocellulose filter which had been pre-soaked in 20X SSC, using the slot blot apparatus of Schleicher and Schuell. 10 μg of DNA was applied per slot. The filters were then baked at 80°C *in vacuo* for 2 hours. Prior to being used for hybridization, the filters were cut into strips and pre-soaked in low-salt hybridization buffer for at least ten minutes. Filters were then incubated in scintillation vials with complete hybridization buffer and labelled nuclear RNA with shaking at 65°C for 2 days. The filters were then washed at 65°C with five changes of 2X SSC for 20 minutes per wash. Finally, the filters were incubated at 37°C in 2X SSC at 10 μg/ml of ribonuclease A for 30 minutes, rinsed briefly in 2X SSC, and exposed to X-ray film with intensifying screens. Hybridization was quantified by cutting out individual strips from the nitrocellulose filters and processing for scintillation counting.

DNA Binding Gel Electrophoresis Assays

Preparation of whole cell extracts and assays for specific DNA-protein interactions were performed as described by Hayes et al. (1987).

RESULTS

Transcription of competence genes

Previous work has shown that there are at least 5 PDGF-inducible "competence" genes.(Cochran et al., 1983; Kelly et al., 1983; Cochran et al., 1984). Two of these are the proto-oncogenes c-*fos* and c-*myc*. Three others, designated JB, JE, and KC, were cloned by differential screening of a cDNA library made from PDGF-treated cells. We have employed *in vitro* runoff transcription assays to determine whether or not the expression of these five competence genes is regulated at the transcriptional level.

Figure 1 shows the results of a runoff transcription experiment performed on confluent, quiescent Balb/c-3T3 cells after a one half and one hour of treatment with v-*sis* conditioned medium (SCM) or cycloheximide, separately and in combination. The conditioned media contains a v-*sis* encoded protein which is very similar to the platelet-derived growth factor in that it has mitogenic activity, binds to the PDGF receptor, and reacts with anti-PDGF antibody (Thiel et al., 1981; Deuel et al., 1983; Owen et al., 1984; Robbins et al., 1983; Johnsson et al., 1985). The results show transcriptional induction of 4 competence genes, JE, JB, KC, and c-*fos*, by 1/2 hour of treatment with v-*sis* conditioned medium. Quiescent cells show low levels of transcription of these genes, except for the c-*fos* and the c-*myc* genes.

The transcriptional activity of the c-*myc* gene in quiescent cells is much more prominent when a c-*myc* genomic clone is used rather than a c-*myc* cDNA clone. This observation suggests that the constitutive transcription from the c-*myc* locus is either coming from transcription in the c-*myc* introns, or is coming from other loci which have sequences which hybridize to the c-*myc* introns.

Treatment with cycloheximide by itself has very little effect on the transcriptional activity of the competence genes. When cells were treated simultaneously with v-*sis*-conditioned medium (SCM) and cycloheximide, the induction of transcription of these genes is not blocked, even though protein synthesis was blocked by more than 90%. Cycloheximide in fact enhances the overall rate of

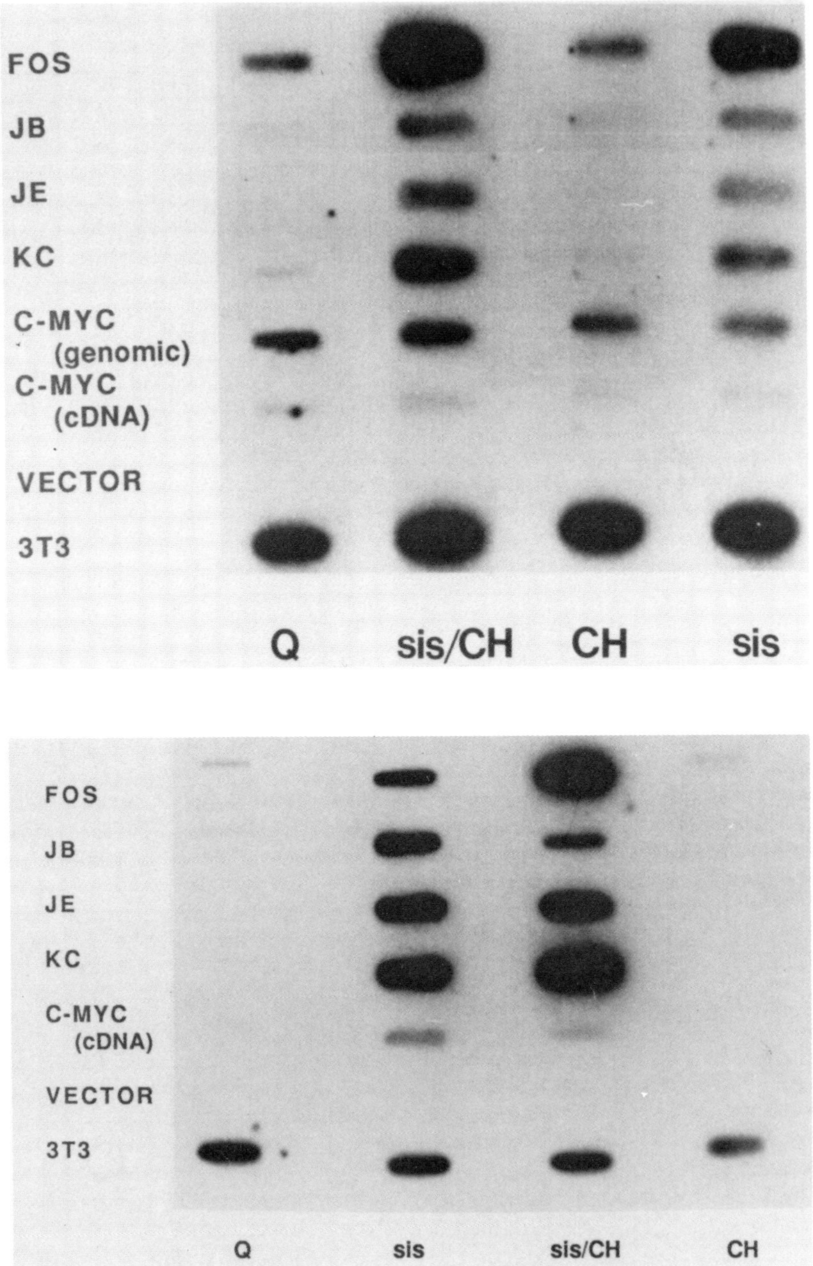

Figure 1. Transcriptional induction of competence genes. *Confluent, quiescent Balb-c-3T3 cells were treated with v-*

sis conditioned media (sis), cycloheximide (10 μg/ml) CH, or both (sis/CH) and runoff transcripts were analyzed as described in Materials and Methods. DNAs were as described in Materials and Methods. The bottom slot labelled 3T3 contains total genomic DNA from mouse 3T3 cells which was used as an internal control for the overall extent of transcription. The top panel shows transcription after one half hour of the indicated treatment. The bottom panel shows transcription after one hour of indicated treatments.

transcription of KC and c-*fos* after SCM stimulation. (See Figure 3) These are the two genes where the superinduction effect is most prominent at the mRNA level. The effect at the transcriptional level is 2 fold for KC and greater than 10 fold for c-*fos*. However, the effect on c-*fos* may be an overestimate in that the time of maximum c-*fos* transcription may be prior to 30 minutes of stimulation (Greenberg et al, 1986). The effect of cycloheximide is not the same for all the competence genes. There is only a very small effect on JE at the transcriptional level, and a diminution of the absolute rates of transcription for c-*myc* and JB.

By one hour after stimulation, the transcriptional activity of JB, JE, and KC is increased and transcriptional induction of the c-*myc* coding exons now can be visualised. This result is consistent with earlier findings which showed that the expression of c-*myc* messenger RNA appears later than does c-*fos* messenger RNA. The transcription of JB, KC and c-myc is maximum at this time. The transcription of JE does not peak until 2 hours. Also by one hour, the effect of cycloheximide treatment is more striking for the c-*fos* gene, as the overall level of transcription of c-*fos* has levelled off in its absence.

Figure 2 shows the transcriptional activity of the competence genes after six hours of SCM treatment. At this time in the cells which were treated with SCM alone, the transcription of all of the competence family genes has begun to return to basal level, with perhaps KC lagging. However, the cells that were treated simultaneously with cycloheximide and SCM continue to show some transcription of all the competence genes. However, this enhancement of transcription by cycloheximide at 6 hours is somewhat

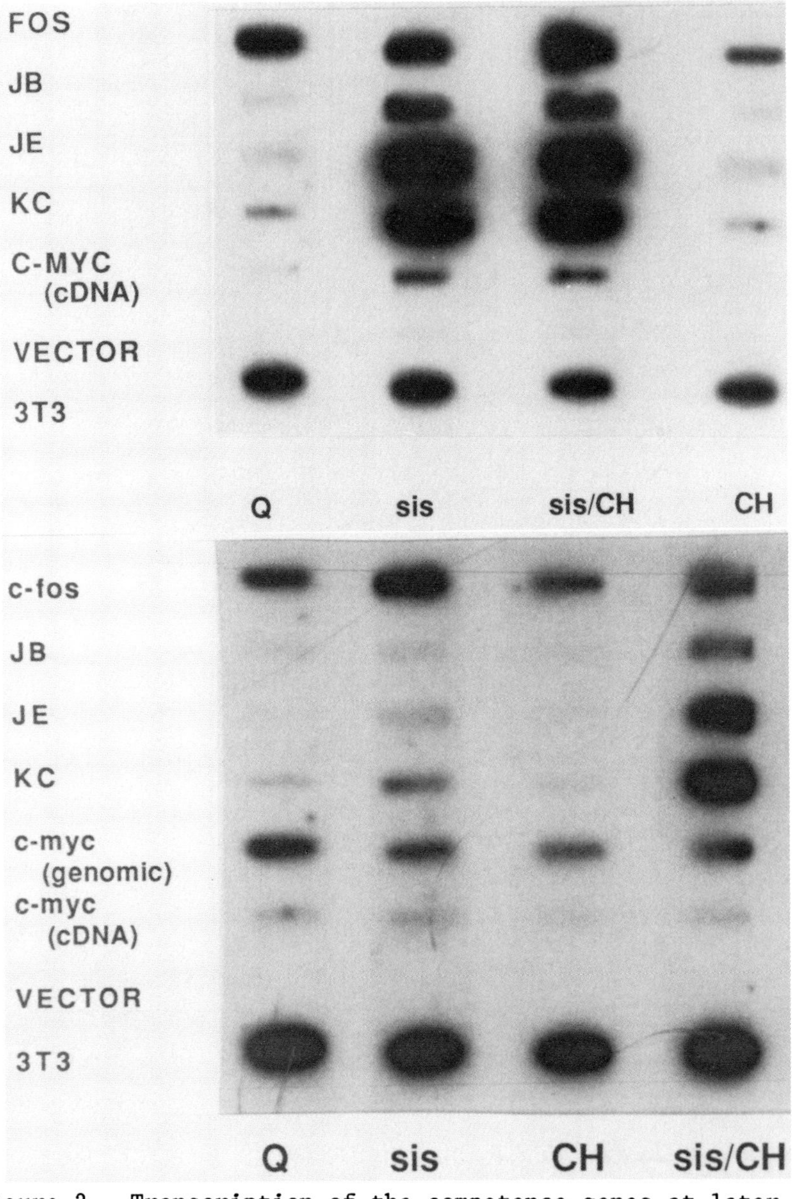

Figure 2. Transcription of the competence genes at later time points. *The experiment was the same as described in Figure 1, except that the time points for the top panel are after 2 hours of treatment, and for the bottom panel are after 6 hours of treatment.*

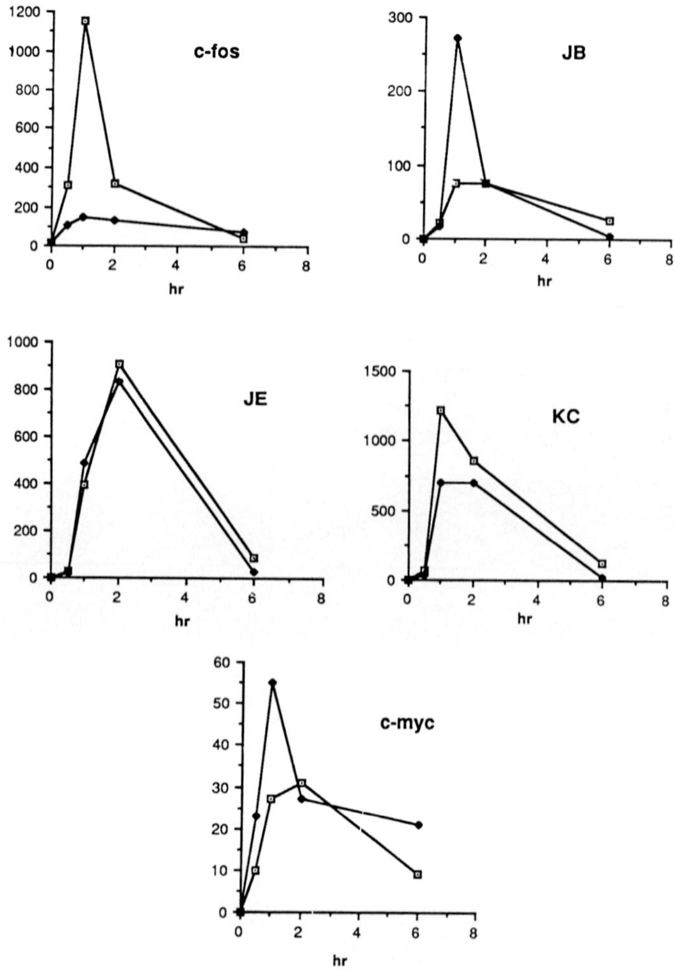

Figure 3. Time course of transcriptional induction of the competence genes. *Data from the experiments shown in Figures 1 and 2 was analyzed by subtracting the number of cpm which hybridized to the vector slot from the number of cpm which hybridized to each test DNA slot and normalizing to an equivalent of 200 cpm hybridized to the 3T3 DNA slot on each filter. The resulting data was plotted for each of the five competence genes as shown above. Filled circles represent transcription after SCM treatment only and open circles show transcription in the presence of SCM plus cycloheximide.*

misleading as the general trend is toward transcriptional shutdown as can be clearly seen from the graphs in Figure 3. The inhibition of protein synthesis appears to only slightly prolong the time course of transcription of the competence genes. Thus, the superinduction effect that has been described for *sis*/PDGF regulated genes has two transcriptional components. One is that certain genes, like the c-*fos* and the KC gene have enhanced overall rates of transcription and the other is that the overall duration of transcription is slightly prolonged. Other data (unpublished observations) also indicate that these messenger RNAs may be stabilized by these treatments indicating that in fact there are three components to the superinduction phenomena.

In the absence of cycloheximide, the maximal transcriptional rate is JE > KC > c-*fos* > JB > c-*myc* with JE transcription being 20 fold higher than that of c-*myc*. In the presence of SCM and cycloheximide, the relative magnitudes of transcription are KC > c-fos > JE > JB > c-*myc*.

Factors which bind upstream of c-*fos*

The experiments described in the previous section make it clear that the competence genes are all induced at the transcriptional level within one hour of addition of v-*sis* conditioned media to quiescent Balb/c-3T3 cells. The question which follows is how is the signal transmitted from the cell surface to individual genes in the nucleus in order to activate transcription. We have chosen to study this question by attempting to identify transacting transcriptional factors which participate in the regulation of competence genes. Such factors should be intermediates in the pathway between the cell membrane and the cell nucleus. The understanding of how these factors and their activities are themselves regulated, should lead to the unraveling of the pathway.

To begin this approach, we have focused on the regulation of the c-*fos* gene. The control elements of this gene have been localized to within the first 400 base pairs upstream of the transcriptional start site (Treisman, 1985; Deschamps et al., 1985). We have employed a DNA-binding gel electrophoresis assay to detect proteins which bind to the c-*fos* control regions. This assay takes advantage of

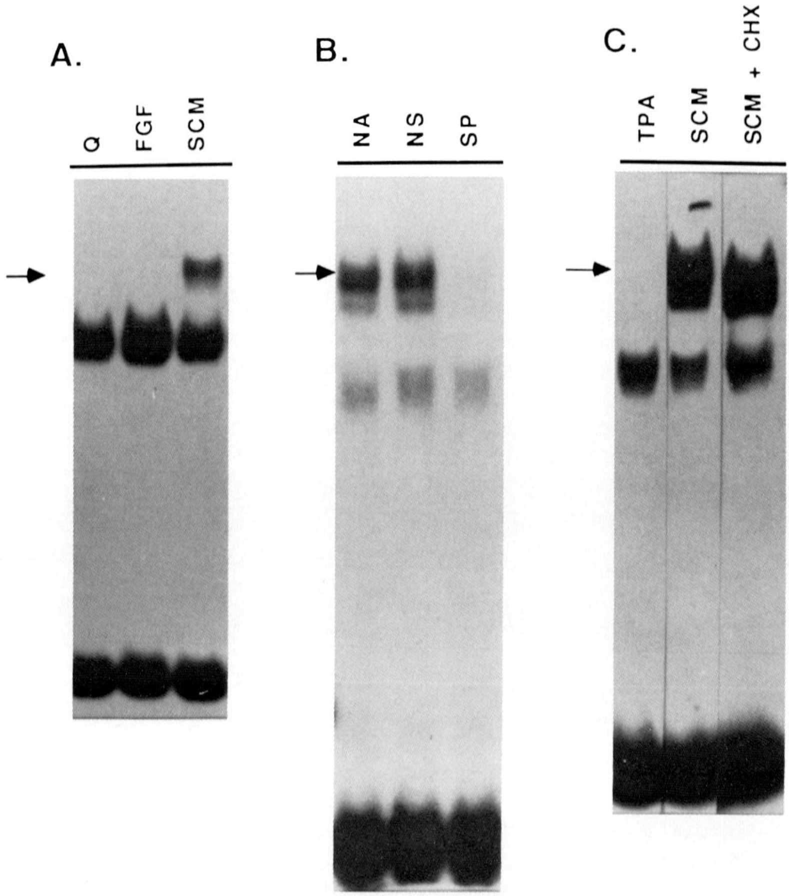

Figure 4. Inducible binding of a cellular factor to the c-*fos* 5′ regulatory region. *Extracts of Balb/c-3T3 cells were analyzed by band shift assays for protein, binding to a double-stranded ^{32}P endlabelled oligonucleotide probe containing residues -351 to -337 on the human c-fos gene. In each case the bottom band is the free probe and the arrow indicates the position of sequence specific binding complexes.*

Panel A) Nuclear extracts were prepared from quiescent Balb-c-3T3 cells (Q) or from cells treated for 30 minutes with 100 ng/ml FGF or with SCM. Each lane contains 12 μg of extract.

Panel B) 10 μg of whole cell extract from cells treated for

30 minutes with SCM were incubated with the following competitor DNAs. NA; No addition, NS, non-specific competition with the 250 base pair fragment from the plasmid vector pGEM1 in a 200-fold excess over labelled probe; SP, specific competition with 250-base pair fragment containing sequences from -361 to -126 of the c-fos gene in a 100-fold molar excess over probe.

Panel C) Confluent, quiescent cells were treated with 100 ng/ml TPA, SCM, or SCM + 10 μg per ml of cycloheximide. 15 μg of whole cell extract was used in each binding reaction. (Figure adapted from Hayes et al., 1987b).

the fact that protein-DNA complexes will migrate more slowly than will free DNA through a low ionic strength polyacrylamide gel (Fried and Crothers, 1981). When sufficient non-specific DNA competitor is incubated with a labelled fragment of interest, and then electrophoresed through such a gel, given DNA complexes will migrate with characteristic mobilities. Figure 4 shows the results of an experiment which demonstrates the DNA-binding activities from Balb/c-3T3 cells that bind to an oligonucleotide which spans the -351 to -337 sequence of the human c-*fos* gene. From Panel A, it can be seen that a new DNA-binding complex appears in cells treated with SCM for 30 minutes which is not present in confluent quiescent cells. The basic fibroblast growth factor fails to induce this new activity, as does the phorbol-ester TPA (Panel C). The experiment in Panel B demonstrates that this new complex is in fact a sequence-specific complex. When an oligonucleotide which spans a c-*fos* (-299/-320) serum response element (Treisman, 1985) was added in 100 fold molar excess to the binding reaction prior to electrophoresis, no effect is seen on the formation of this inducible DNA-binding complex. If, however, a 100-fold molar excess of an unlabelled c-fos DNA fragment which spans the -351/-337 sequence of the labeled oligonucleotide probe is added to the binding reaction, this inducible complex is entirely competed away. This demonstrates that the inducible factor is binding in a sequence-specific manner, and is also a factor which is distinguishable from the one which other laboratories have identified as binding to a region of dyad symmetry centered around -310 of the *fos* gene (Treisman, 1987; Gilman et al., 1986; Prywes and Roeder, 1986). The experiments in Panel C demonstrate that this DNA-binding factor is inducible even in the absence of new protein synthesis. Thus, the induction of this factor by SCM is not blocked by the addition of cycloheximide to the cultures. In fact, there is some superinduction of the factor with cycloheximide, but it is only 2-fold, whereas the superinduction of the gene at the transcriptional level is approximately a 10-fold effect. The results of this experiment make it seem likely that the factor exists in quiescent cells in an inactive form and has its DNA-binding activity activated only after treatment of the cells with SCM, and that this activation does not require new protein synthesis. Thus, this factor has several of the characteristics necessary for it to be an intermediate in signal transduction pathways.

DISCUSSION

The experiments described in this paper have demonstrated that 5 competence genes are rapidly regulated at the transcriptional level by v-*sis* conditioned media. We have obtained similar results using purified v-*sis* protein. The time course and intensity of the transcriptional induction varies for individual genes, but there is no question that there is transcriptional induction for all of the genes including the c-*myc* gene. There appears to be some type of constitutive transcriptional activity originating either from within the c-*myc* introns, or from genes which have sequences which are closely related to those of the c-*myc* introns. However, when a *myc* cDNA probe is used in a runoff transcription assay, induction at the transcriptional level is clear. It may be that there is post-transcriptional control of this gene as well, and the experiments described here do not rule out that possibility. The superinduction effect which has been observed for some of the competence genes after induction in the presence of protein synthesis inhibitors appears to be mediated in two ways at the transcriptional level. Primarily, there is an overall increase in the level of transcription, which is especially prominent for the c-*fos* and the KC genes, and secondarily, there is a prolongation of the time course of transcription. For JB, JE, and c-*myc*, cycloheximide does not superinduce the rate of transcription.

Using a DNA-binding gel electrophoresis assay, we have identified a DNA-binding factor whose activity is induced by treatment of quiescent cells with v-*sis* conditioned media. This induction happens within 20 minutes of treatment of the cells and the induced activity binds approximately 25 bp upstream of another factor that binds to a dyad symmetry element centered at -310 of human c-fos. This factor is a potential intermediate in the signalling processes which connect the cell surface to the cell nucleus. An interesting next question to be asked about this factor is whether or not it binds to the regulatory sequences of other competence genes. Preliminary results indicate potential binding sites for this factor upstream of the c-*myc* gene. However, if this is the case, it still does not explain the difference in the time course and magnitude of the induction of the c-*myc* and the c-*fos* genes. Current work in the laboratory is devoted toward purifying this factor and analyzing its role in the regulation of expression of the c-*fos* and the c-*myc* genes.

ACKNOWLEDGEMENTS

This work was supported by grants to B.H.C. from the American Cancer Society and the National Institute of Health. T.E.H. is an American Cancer Society fellow. B.H.C. is a Rita Allen Scholar and Searle Scholar. We thank Karen Traite for help in preparing the manuscript.

REFERENCES

Armelin HA, Armelin MCS, Kelly K, Stewart T, Leder P, Cochran BH, Stiles CD (1984). A functional role for c-*myc* in the mitogenic response to platelet-derived growth factor. Nature 310:655-660.

Clemens MJ, McNurlan MA (1985). Regulation of cell proliferation and differentiation by interferons. Biochem J 226:345-360.

Cochran BH (1985). The molecular action of platelet-derived growth factor. Advances in Cancer Research 45:183-216.

Cochran BH, Reffel AC, Stiles CD (1983). Molecular cloning of gene sequences regulated by platelet-derived growth factor. Cell 33:939-947.

Cochran BH, Zullo J, Verma IM, Stiles CD (1984). Expression of the c-*fos* gene and of any *fos*-related gene is stimulated by platelet-derived growth factor. Science 226:1080-1082.

Deschamps J, Meijlink F, Verma IM (1985). Identification of a transcriptional enhancer element upstream from the proto-oncogene *fos*. Science 230:1174-1177.

Deuel TF, Huang JS, Huang SS, Stroobant P, Waterfield MD (1983). Expression of a platelet-derived growth factor-like protein in simian sarcoma virus transformed cells. Science 221:1348-1350.

Doolittle RF, Hunkapiller MW, Hood LE, Devare SG, Robbins KC, Aaronson SA, Antoniades HN (1983). Simian sarcoma virus oncogene, v-*sis*, is derived form the gene (or genes) encoding a platelet-derived growth factor. Science 221:275-277.

Downward J, Yarden Y, Mayes E, Scrace G, Totty N, Stockwell P, Ullrich A, et al. (1984). Close similarity of epidermal growth factor receptor and v-*erb-B* oncogene protein sequences. Nature 307:521-527.

Fried M, Crothers DM (1981). Equilibria and kinetics of lac repressor-operator interactions by polyacrylamide gel electrophoresis. Nucleic Acids Research 9:6505-6525.

Gilman MZ, Wilson RN, Weinberg RA (1986). Multiple protein binding sites in the 5'-flanking region regulate c-*fos* expression. Mol Cell Biol 6:4305-4316.

Greenberg ME, Ziff EB (1984). Stimulation of 3T3 cells induces transcription of the c-*fos* proto-oncogene. Nature 311:433-442.

Greenberg ME, Hermanowski AL, Ziff EB (1986). Effect of protein synthesis inhibitors on growth factor activation of c-*fos*, c-*myc* and actin gene transcriptiuon. Mol Cell Biol 6:1050-1057.

Hayes TE, Kitchen AM, Cochran BH (1987a). Inducible binding of a factor of the c-*fos* regulatory region. Proc Natl Acad Sci USA 84:1272-1276.

Hayes TE, Kitchen AM, Cochran BH (1987b). A rapidly inducible DNA-binding activity which binds upstream of the c-*fos* proto-oncogene. J Cell Physiol (in press)

Hurley JB, Simon MI, Teplow DB, Robishaw JD, Gilman AG (1984). Homologies between signal transducing G proteins and *ras* gene products. Science 226:860-862.

Johnsson A, Betsholtz C, Helm KVder, Heldin C-H, Westermark B (1985). Platelet-derived growth factor agonist activity of a secreted form of the v-sis oncogene product. Proc Natl Acad Sci USA 82:1721-1725.

Kaczmarek L, Hyland JK, Watt R, Rosenberg M, Baserga R (1985).Microinjected c-*myc* as a competence factor.

Science
228:1313-1315.

Kelly K, Cochran BH, Stiles CD, Leder P (1983). Cell-specificregulation of the c-*myc* gene by lymphocyte mitogens and platelet-derived growth factor. Cell 35:603-610.

Massague J (1987). The TGF-β family of growth and differentiation factors. Cell 49:437-438.

Miller AD, Curran T, Verma IM (1984). c-*fos* protein can induce cellular transformation: a novel mechanism of activation of a cellular oncogene. Cell 36:51-60.

Nishikura K, Murray JM (1987). Antisense RNA of proto-oncogene c-*fos* blocks renewed growth of quiescent 3T3 cells. Mol Cell Biol 7:639-649.

Owen AJ, Pantazis P, Antoniades HN (1984). Simian Sarcoma virus-transformed cells secrete a mitogen identical to platelet-derived growth factor. Science 225:54-56.

Pledger WJ, Stiles CD, Antoniades HN, Scher CD (1978). An ordered sequence of events occurs before Balb/c-3T3 cells become committed to DNA synthesis. Proc Natl Acad Sci USA 75:2839-2843.

Prywes R, Roeder RG (1986). Inducible binding of a factor to the c-*fos* enhancer. Cell 777-784.

Robbins KC, Antoniades HN, Devare SG, Hunkapillar MW, Aaronson SA (1983). Structural and immunological similarities between derived growth factor. Nature 305:605-608.

Scher CD, Pledger WJ, Martin P, Antoniades H, Stiles CD (1978). Transforming viruses directly reduce the cellular growth requirement for a platelet derived growth factor. J Cell Physiol 97:371-380.

Stanton LW, Watt R, Marcu K (1983). Translocation, breakage and truncated transcripts of c-*myc* oncogene in murine plasmacytomas. Nature 303:401-406.

Stiles CD, Capone GT, Scher CD, Antoniades HN, Van Wyck JJ, Pledger WJ (1979). Dual control of cell growth by somatomedins and platelet-derived growth factor. Proc. Natl. Acad. Sci. 76:1279-1283.

Thiel H-J, Matthews TJ, Weinhold KJ, Broughton EM (1981). Identification of a 20,000-Dalton protein in SSV-transformed nonproducer cells. J Virol 115:401-405.

Treisman R (1985). Transient accumulation of c-*fos* RNA following serum stimulation requires a conserved 5′ element and c-*fos* 3′ sequences. Cell 42:889-902.

Treisman R (1986). Identification of a protein-binding site that mediates transcriptional response of the c-*fos* gene to serum factors. Cell 46:567-574.

Index